Pest Management in Vegetables

Pest Management in Vegetables

Part - I

Dr. I.M. Khan

AGROTECH PRESS

JAIPUR ▪ NEW DELHI

AGROTECH PRESS

Regd. Off. "Bony Residency", Gate No. 2
Opp. Tilak Public School, Vishweswariya Nagar,
Gopalpura Road, Jaipur - 302018, India
Ph: 0141 - 2761280, Telefax: 2761381
E-mail : agrotechpress@gmail.com

Delhi Office: 102, 1st Floor, Satyam House,
4327/3, Ansari Road, Darya Ganj,
New Delhi - 110002, India
Ph: 011-45652440
visit us: www.agrotechpress.com

Distributed By
KEYA INTERNATIONAL PRESS LLP
E-mail: keyainternationalpressllp@gmail.com

Published in India, 2024

ISBN : 9789384568436 (2 Vols. Set)

Typeset by Thomson Press

Printed and bound in India by Thomson Press Ltd.,
New Delhi

Contents

Preface

Vegetables are important components of daily diet throughout the world. Vegetables provide nutritional security in addition to food security. Their production extends to billions of tons and source of income for a large population. Yield and quality are central to sustainable vegetable production. If not properly managed, pests and diseases can dramatically reduce crop yield, quality and subsequent returns. A lot of time, money and natural resources are invested into growing vegetables. Even in the best-managed vegetable gardens - ones with soil rich in compost, and a diversity of plants to encourage natural predators - certain pests will occasionally get out of hand. Good pest and disease management can protect this investment from avoidable losses.

Limited tools to combat these problems is one of the major contributing factors to the high risk associated with vegetable production in general. Consequently, high level of management is required to successfully control pests' problems. Successful production of organically grown vegetables requires an even higher level of management for profitability to occur because even fewer control tools are available to producers using this production system. Organic producers have to rely heavily upon them minor details of vegetable culture in order to avoid problems. Pest control is often a result of the cumulative effects of many production practices. Emphasis is placed on prevention rather than control. This often is difficult to achieve with current technology. Therefore, a grower must develop pest control strategies. To do so a grower must have an understanding of the pest and the beneficials that usually occur in a given crop, and, he should also have the ability to anticipate pest problems and manipulate practices that minimize pests and maximize beneficial. Most of these tactics are used as preventive measures in an attempt to help keep pests below critical levels. Even after preventive measures are taken, certain conditions can still allow a pest population to build up to injurious levels. If pests do build up, then a rescue treatment can be used to reduce the population of the pest and rescue the crop. At the heart of pest management is the understanding that most crops can tolerate a certain amount of pest damage, thus a rescue treatment is not recommended until the pest population reaches a critical level that is referred to as a threshold.

Knowledge of which specific pests are present in the garden, along with background information about each pest's habits and life cycle, is the gardener's key to planning the most appropriate management strategy. Most insects found in gardens are not pests. Many are beneficial, preying on pests or performing other useful tasks. Correctly identifying the insects and other pests that attack vegetables is the first step toward controlling these pests effectively. One of the most important strategies for dealing with insects is to learn about insect life cycles, behaviors, habitats, and diets, and to recognize which are pests and which are actually lending you a helping hand. The pest descriptions in this book are designed to help the commercial vegetable grower and home gardener recognize the insects found on vegetables. Information on insect life cycles and habits is included to make pest control recommendations more understandable and usable.

Today, pests and diseases are managed using a more integrated approach. These approaches bring together the best mix of chemical and biological controls and cultural practices, to manage pests and diseases. They don't discard traditional chemical treatments and local knowledge, but integrate them into a sustainable system. Cultural practices are among the oldest techniques used for pest suppression, and many of the preventative practices used in conventional and organic farming today have their roots in traditional agriculture. Because only a limited range of suppressive pest control tactics are available to organic growers, knowledge-intensive cultural practices form the basis of organic pest management programs. Cultural, mechanical, physical, genetic, and biological controls help prevent severe pest problems, while pesticides are used when additional control measures are required. In a least-toxic pest control strategy a good step is to employ barriers, such as row covers, to exclude pests altogether. Using a pesticide, any pesticide, is always the measure of last resort. Gardeners today have at their disposal a handful of effective and safe pesticides.

Written with expertise from entomology, plant pathology, plant breeding, plant physiology, biochemistry, and extension education covering externalities of pesticide use, pest outbreaks, threshold theory, host plant resistance, crop plant manipulation, biological control, behavioral-modifying techniques, botanicals, and non-pesticide pest management, the book is an excellent source of advanced study material for academics, researchers and students, elucidating the underlying concepts of the pest management development process.

Dr. I.M. Khan

1

Introduction

In culinary terms, a vegetable is an edible plant or its part, intended for cooking or eating raw. In biological terms, "vegetable" designates members of the plant kingdom.

The non-biological definition of a vegetable is largely based on culinary and cultural tradition. Apart from vegetables, other main types of plant food are fruits, grains and nuts. Vegetables are most often consumed as salads or cooked in savory or salty dishes, while culinary fruits are usually sweet and used for desserts, but it is not the universal rule. Therefore, the division is somewhat arbitrary, based on cultural views. For example, some people consider mushrooms to be vegetables even though they are not biologically plants, while others consider them a separate food category; some cultures group potatoes with cereal products such as noodles or rice, while most English speakers would consider them vegetables.

Some vegetables can be consumed raw, while some, such as cassava, must be cooked to destroy certain natural toxins or microbes in order to be edible. A number of processed food items available on the market contain vegetable ingredients and can be referred to as "vegetable derived" products. These products may or may not maintain the nutritional integrity of the vegetable used to produce them.

The word *vegetable* was first recorded in English in the early 15th century from Old French, and was in origin applied to any plant. This is still the sense of the adjective "vegetable" in biological context. The word is taken from Medieval Latin *vegetabilis* "growing, flourishing" (i.e. of a plant), a semantic change from a Late Latin meaning "to be enlivening, quickening", a derivation of the verb *vegetare* "enliven", which is derived from *vegetus* "to be alive, active" (related to *vigor*), in reference to the process of a plant growing.

In 1767, the meaning of the term "vegetable" was specified to mean "plant cultivated for food, edible herb or root." The year 1955 noted the first use of the shortened, slang term "veggie". As an adjective, the word *vegetable* is used in scientific and technical contexts with a different and much broader meaning, namely of "related to plants" in general, edible or not — as in *vegetable matter, vegetable kingdom, vegetable origin*, etc. The meaning of "vegetable" as "plant grown for food" was not established until the 18th century.

Examples of Different Parts of Plants Used as Vegetables

Figure: *Farmers' market showing vegetables for sale near the Potala palace in Lhasa, Tibet.*

Figure: *Melon Yard at Heligan Cornwall, England, UK*

The list of food items called "vegetable" is quite long, and includes many different parts of plants:

- Flower bud broccoli, cauliflower, globe artichokes, capers
- Leaves kale, collard greens, spinach, arugula, beet greens, bok choy, chard, choi sum, turnip greens, endive, lettuce, mustard greens, watercress, garlic chives, gai lan
- Leaf sheaths leeks
- Buds Brussels sprouts
- Stem Kohlrabi and galangal
- Rhizome Ginger
- Stems of leaves celery, rhubarb, cardoon, Chinese celery
- Stem shoots asparagus, bamboo shoots
- Tubers potatoes, Jerusalem artichokes, sweet potatoes, taro, and yams
- Whole-plant sprouts soybean, mung beans, urad, and alfalfa
- Roots carrots, parsnips, beets, radishes, rutabagas, turnips, and burdocks
- Bulbs onions, shallots, garlic
- Fruits in the botanical sense, but used as vegetables tomatoes, cucumbers, squash, zucchinis, pumpkins, peppers, eggplant, tomatillos, chayote, okra, breadfruit, avocado, pods, seeds such as corn, green beans and snow peas.

Nutrition

Vegetables are eaten in a variety of ways, as part of main meals and as snacks. The nutritional content of vegetables varies considerably, though generally they contain little protein or fat, and varying proportions of vitamins such as Vitamin A, Vitamin K and Vitamin B6, provitamins, dietary minerals and carbohydrates.

Vegetables contain a great variety of other phytochemicals, some of which have been claimed to have antioxidant, antibacterial, antifungal, antiviral and anticarcinogenic properties.

Some vegetables also contain fibre, important for gastrointestinal function. Vegetables contain important nutrients necessary for healthy hair and skin as well. A person who refrains from dairy and meat products, and eats only plants (including vegetables) is known as a vegan.

However, vegetables often also contain toxins and antinutrients such as α-solanine, α-chaconine, enzyme inhibitors (of cholinesterase, protease, amylase, etc.), cyanide and cyanide precursors, oxalic acid, and more. Depending on the concentration, such compounds may reduce the edibility, nutritional value, and health benefits of dietary vegetables. Cooking and/or other processing may be necessary to eliminate or reduce them.

Diets containing recommended amounts of fruits and vegetables may help lower the risk of heart diseases and type 2 diabetes. These diets may also protect against some cancers and decrease bone loss. The potassium provided by both fruits and vegetables may help prevent the formation of kidney stones.

Safety

Even in the best-managed vegetable farms—ones with soil rich in compost, and a diversity of plants to encourage natural predators—certain pests will occasionally get out of hand. There is a nugget of truth in the old maxim that insects are most likely to gang up on plants that are already unhealthy because of a soil nutrient imbalance or drought. This is the case, for examply, with some aphid outbreaks. Unfortunately, the maxim doesn't come close to explaining all our pest problems.

For instance, vegetables are bred largely for yield and flavour, often at the expense of natural resistance to pests. Furthermore, all vegetables are tender and nutritious, and this fact is not lost on a wide array of insects.

With cabbageworms, hornworms, bean beetles, and Colorado potato beetles, the better you've made the soil, the more they like your vegetables. If these or similar insects are in your neighbourhood and you are growing their favourite crops, you are almost certain to have a pest outbreak.

Integrated Pest Management

Vegetable garden pest control begins with basic good gardening common sense, such as choosing varieties that are resistant to pests in your region, preparing the soil well and providing regular irrigation.

It helps to have in your garden a diversity of plants and habitats. Water, even a very small pond, is attractive to many insects and other creatures. Likewise, an abundance of flowering, nectar-bearing plants will encourage and sustain parasitic and predatory insects.

The next step in a least-toxic pest control strategy is to employ barriers, such as row covers, to exclude pests altogether. Using a pesticide, any pesticide, is always the measure of last resort. You spray or dust late in the game, when the pest insect is clearly way out of control and an important crop is at risk.

When a pest first arrives, or when prior experience tells you it soon will, the best approach is to develop a strategy of control. Integrated pest management (IPM) is a pest problem-solving process that includes considerations such as pesticide resistance, natural biological controls and pollution, in addition to problems caused by the pest.

IPM integrates many pest-control methods and minimizes insecticide use, particularly of the more toxic, broad-spectrum kinds.

When a problem does occur, it is essential to correctly identify the cause. The beetle you see near a hole in a leaf may be a predator. But if it is damaging your plants, simply pick it off. Also consider that doing nothing at all—letting nature take its course—is often the best approach. Always use simple, noninvasive remedies first.

Sometimes pest problems are not adequately managed by natural, cultural or mechanical control methods. Insecticides are often the only control option that remains. The prime factors in determining pesticide safety are:

- How specific it is to particular insects?
- How toxic it is to humans and nontarget organisms?
- How quickly does it degrade.

Remember, in a vegetable garden it is virtually impossible to spray just the one thing you want to spray. Other crops (perhaps ready to pick) are always nearby, so you want to stay away from insecticides that don't break down quickly.

Choose an insecticide that is as specific to the pest at hand as possible and then use as little as possible. If only one spray will do the job, use only one spray. For the long-term health of your garden, the less spray you use, the better.

Remember, too, that just because an insecticide has a botanical origin or is considered acceptable to organic gardeners, it still contains a toxin and is not automatically safe for humans.

Vegetable Pest Remedies

Gardeners today have at their disposal a handful of effective and safe pesticides. When you have to spray to save your crop, here are the insecticides to consider using, with their characteristics, positive and negative.

Bt.: The bacterium *Bacillus thuringiensis* (Bt) was identified in 1911 by the biologist E. Berliner, who found it infecting pupae of the Mediterranean flour moth and other insect larvae living in grain warehouses in the German town of Thuringia. It wasn't until the 1960s, however, that entomologists learned how to make it into a powerful and very pest-specific insecticide.

Advantages of Bt include safety—it is essentially nontoxic to humans, other mammals and birds. The label specifies no waiting period between application and harvest. It is also highly selective so is easily incorporated with existing natural controls. A limitation of Bt is its slow action. After pests consume it, their feeding slows down. But their death won't occur for two to five days.

Bt is also perishable. Most formulations are less effective after a few years of storage. Bt exists naturally in most soils. Different strains of Bt occur that produce protein crystals toxic to certain insects. The strain for most caterpillars is *B. t. kurstaki.* Commercially prepared Bt spray or powder has no effect on adult butterflies or moths. Remember, however, that not all caterpillars are pests.

Strains of Bt have been developed for a few other pests. Some leaf-feeding beetles (including Colorado potato beetles) are susceptible to *B. t. tenebrionis* (also known as San Diego strain and M-1), for example. Because Bt is a near-perfect insecticide, there is danger of overuse. Any overused insecticide will gradually become less effective as insects evolve defences to it. Some insect pests, such as the diamondback moth and Indian meal moth, were once susceptible and are now at least partially immune to Bt.

Bt is in the news because plant scientists have learned how to insert its genes into corn plants making the entire plant toxic to corn earworm. While Bt is favoured and accepted by all organic gardeners, genetically engineering Bt toxins into other plants is not.

Diatomaceous Earth (DE): This is a powder-like dust made of the silicate skeletons of tiny marine creatures called diatoms. Millions of years ago as they died, their skeletons gradually accumulated into deep layers that are mined today from deposits where oceans or large lakes once covered the land. DE acts like ground glass, cutting into the waxy coat of insects and causing them to dry out and die. It is not toxic if eaten, but is irritating if inhaled.

DE is available in two different forms. One form is used primarily in swimming pool filters. It is not an effective insecticide and is dangerous to inhale (it can cause a lung disease called silicosis). In your garden, use only the natural grade of DE. Still, it is wise to wear goggles and a dust mask during application. Dust DE onto leaves and stems to control pests such as aphids, Colorado potato beetle, immature forms of squash bug, Mexican bean beetle or whitefly. Or spread it as a barrier to slugs and snails. It works best in dry situations. It is not selective and kills spiders and beneficials as well as pests. Don't overuse it.

Particle Film Dusts: Dusts are among the oldest types of insecticides, but the newest kinds are very carefully designed for maximum efficiency. Only one type is currently available to home gardeners, Surround At Home. It's made of an engineered kaolin clay. Applied with a pressurized sprayer, it coats plant leaves making them less familiar and less comfortable to pests. Since these are not insecticides, there is no danger of a pest developing resistance. And the kaolin, a common ingredient in medicines, is non toxic.

Horticultural Oils: These are most often highly refined extracts of crude oil. (Some vegetable oils, such as cottonseed and soybean oil, are also sometimes used.) They are increasingly recommended for vegetable garden pest control because they present few risks to either gardeners or desirable species and integrate well with natural biological controls. Also, oils dissipate quickly through evaporation, leaving little residue.

Oils kill insects by plugging the pores through which they breathe. Oils can damage plants if applied at excessive rates or on particularly hot (above 100°F) or cold (below 40°F) days.

Spray oils in vegetable gardens to kill aphids, leafhoppers, spider mites and whiteflies. A few drops of oil in the ear tips of corn controls corn earworm.

Insecticidal Soaps: These are specific fatty acids that have been found by experiment to be toxic to pests, primarily soft-bodied insects such as aphids, mealybugs, spider mites and whiteflies. Surprisingly, adult Japanese beetles are also susceptible. Most nontarget insects are unaffected, and toxicity to animals is nonexistent. Soap insecticides act fast and leave no residue. You can use them on vegetables up to the moment of harvest. Advantages of soaps to home gardeners include safety to both the applicator and nontarget insects. They a selective, so are easily incorporated with other, natural biological controls. Some plants, such as peas, are readily burned by soaps, and their effectiveness is greatly reduced if mixed with hard water.

Don't use liquid dishwashing detergents or hand soaps. Though many will kill insects, they might hurt the plants, too. Some fatty acids are toxic to plants, the reason a soap-based weed killer is now available.

Dish soap manufacturers change the oils used in their formulations regularly, based on cost and availability, so the brand that worked fine for your neighbour last year might severely stunt your Brussels sprouts this year.

Neem: This is an extract derived from the crushed seeds of the tropical neem tree (*Azadirachta indica*). Though intensely studied for many years now, it is still a new botanical insecticide. The primary active ingredient is the compound azadirachtin, although the oils and other ingredients also have some insecticidal effect. According to EPA regulations, neem is exempt from food crop tolerances because it is considered nontoxic.

Neem works both as an insecticide and as an antifeedant. It kills insects in the juvenile stage by thwarting their development, and is most effective against aphids, thrips and whiteflies. There is no quick knockdown with neem, but a week or so after application, you'll notice a steady decline in the number of pests. It is not effective against adult insects (though it may interfere with egg production), and has little impact on beneficial insects.

As an antifeedant, neem is effective against Japanese beetles. Apply neem before the beetles appear and reapply after rainfall. Once beetle numbers build up on the plant, neem no longer discourages them.

Neem sprays degrade very quickly in water. Mix only the amount you need and apply all of it immediately. On the plant, neem retains its activity against juvenile insects pests for about one week.

Pyrethrins: Derived from the painted daisy, *Chrysanthemum cinerariifolium*, pyrethrins are considered one of the most important natural insecticides. When you must use a broad-spectrum insecticide in the vegetable garden or lose the crop, they are one of the best choices. Of low toxicity to mammals, they kill insects quickly. In sunlight they break down and are nontoxic within a day or less. For best results, apply them in the late afternoon or evening. Use pyrethrins for the hard-to-kill pests such as beetles, squash bugs and tarnished plant bugs.

The terminology can be confusing. Pyrethrum, discovered around 1800 in the Transcaucasian region of Asia, is the ground-up flowers of the daisy. Pyrethrins (almost always plural) are the insecticidal components of the flowers. Pyrethroids, such as cypermethrin, permethrin and resmethrin, are synthetic compounds that resemble pyrethrins. They are more toxic and more persistent than pyrethrins, so are much more toxic to beneficials. Though increasingly popular in commercial agriculture, home gardeners should avoid them.

Often, pure pyrethrins only stun insects. This is why they are often combined with synergists like piperonyl butoxide, chemicals that enhance the effectiveness of the active ingredients, thus enabling formulations with less pyrethrin to kill insects.

Eight Vegetable Garden Insecticides

Here are the insecticides you are most likely to need in a vegetable garden. Note that many others are available and sometimes recommended, but these are the safest and most effective for most home gardeners.

Name	***Use against***	***How it works; how long it lasts***	***How to use it***
Bt. Toxic primarily to caterpillars	Cabbageworms and cutworms. Use *B. t. tenebrionis* for Colorado potato beetle.	Bacterial toxin; causes caterpillar death usually within 24 hours. Dissipates in 2 days or less.	Available as spray or dust. Apply late afternoon and reapply after rain. Mix with insecticidal soap for better coverage.
Diatomaceous earth (DE). Toxic primarily to soft-bodied insects, snails and slugs	Discourages aphids, juivenile Colorado potato and bean beetles, leafhoppers, and thrips.	Sharp-edged diatom skeletons scratch insect exteriors, causing them to dry out.	More effective combined with pyrethrins. Use natural grade DE, not the kind used in swimming pool filters. Wear dust mask.
Particle film dust. Coats leaves and repels insects	Japanese beetles, tarnished plant bugs, and Colorado potato beetles.	Specially refined kaolin clay uses microscopic particulates to form barrier film on leaves.	Apply with pressure sprayer. Mix with Bt or pyrethrins.
Horticultural oils. Toxic to many insects and mites	Kills aphids, corn earworms, leafhoppers, spider mites, whiteflies.	Kills pests by suffocating them.	Use highly refined summer oils. Do not apply to drought-stressed plants, or on hot, cold, or very humid days.
Insecticidal soap. Toxic mostly to soft-bodied insects	Kills aphids, earwigs, grasshoppers, Japanese beetle adults, leafhoppers, spider mites, whiteflies.	One of the safest insecticides. Fatty acids destroy the cellular membrane of insects on contact.	Mix with warm soft water and be sure to cover both sides of leaves. Can burn leaves during hot weather.
Neem. Toxic to juvenile forms of some pests; also a repellent	Kills juvenile aphids and Colorado potato beetle. Repels whiteflies, Japanese and Colorado beetle adults.	Affects growth hormones of some insects causing them to stop feeding. Effect lasts about 1 week.	Apply liquid spray morning or evening with humidity is high.
Pyrethrins Toxic to a broad spectrum of pests	Controls most vegetable pests included flea, potato, and bean beetles.	A nerve toxin, often combined or DE. Degrades rapidly.	Apply dust during cloudy weather or early evening.

Vegetable Farming

Vegetable farming is the growing of vegetables for human consumption. Traditionally it was done in the soil in small rows or blocks, often primarily for consumption on the farm, with the excess sold in nearby towns. Later, farms on the edge of large communities could specialise in vegetable production, with the short distance allowing the farmer to get his produce to market while still fresh. The three sisters method used by Native Americans (specifically the Haudenosaunee/Iroquois) grew squash, beans and corn together so that the plants enhanced each other's growth.

Planting in long rows allows machinery to cultivate the fields, increasing efficiency and output; however, the diversity of vegetable crops requires a number of techniques to be used to optimize the growth of each type of plant. Some farms, therefore, specialise in one vegetable; others grow a large variety. Due to the needs to market vegetables while fresh, vegetable gardening has high labour demands. Some farms avoid this by running u-pick operations where the customers pick their own produce. The development of ripening technologies and refrigeration has reduced the problems with getting produce to market in good condition. Over the past 100 years a new technique has emerged—raised bed gardening, which has increased yields from small plots of soil without the need for commercial, energy-intensive fertilizers. Modern hydroponic farming produces very high yields in greenhouses without using any soil.

Several economic models exist for vegetable farms: farms may grow large quantities of a few vegetables and sell them in bulk to major markets or middlemen, which requires large growing operations; farms may produce for local customers, which requires a larger distribution effort; farms may produce a variety of vegetables for sale through on-farm stalls, local farmer's markets, u-pick operations. This is quite different from commodity farm products like wheat and maize which do not have the ripeness problems and are sold off in bulk to the local granary. Large cities often have a central produce market which handles vegetables in a commodity-like manner, and manages distribution to most supermarkets and restaurants.

In America, vegetable farms are in some regions known as truck farms; "truck" is a noun for which its more common meaning overshadows its historically separate use as a term for "vegetables grown for market". Such farms are sometimes called muck farms, after the dark black soil in which vegetables grow well.

Kitchen Garden

The traditional kitchen garden, also known as a potager (in French, *jardin potager*) or in Scotland a kailyaird, is a space separate from the rest of the residential garden – the ornamental plants and lawn areas. Most vegetable gardens are still miniature versions of old family farm plots, but the kitchen garden is different not only in its history, but also its design.

The kitchen garden may serve as the central feature of an ornamental, all-season landscape, or it may be little more than a humble vegetable plot. It is a source of herbs, vegetables and fruits, but it is often also a structured garden space with a design based on repetitive geometric patterns. The kitchen garden has year-round visual appeal and can incorporate permanent perennials or woody shrub plantings around (or among) the annuals.

Potager Garden

A potager is a French term for an ornamental vegetable or kitchen garden. The historical design precedent is from the Gardens of the French Renaissance and Baroque Garden à la française eras. Often flowers (edible and non-edible) and herbs are planted with the vegetables to enhance the garden's beauty. The goal is to make the function of providing food aesthetically pleasing.

Plants are chosen as much for their functionality as for their colour and form. Many are trained to grow upward. A well-designed potager can provide food, as well as cut flowers and herbs for the home with very little maintenance. Potagers can disguise their function of providing for a home in a wide array of forms—from the carefree style of the cottage garden to the formality of a knot garden.

Vegetable Garden

A vegetable garden (also known as a vegetable patch or vegetable plot) is a garden that exists to grow vegetables and other plants useful for human consumption, in contrast to a flower garden that exists for aesthetic purposes. It is a small-scale form of vegetable growing. A vegetable garden typically includes a compost heap, and several plots or divided areas of land, intended to grow one or two types of plant in each plot. Plots may also be divided into rows with an assortment of vegetables grown in the different rows. It is usually located to the rear of a property in the back garden or back yard. Many families have home kitchen and vegetable gardens that they use to produce food. In World War II, many people had a garden called a "victory garden" which provided food and thus freed resources for the war effort.

With worsening economic conditions and increased interest in organic and sustainable living, many people are turning to vegetable gardening as a supplement to their family's diet. Food grown in the back yard consumes little if any fuel for shipping or maintenance, and the grower can be sure of what exactly was used to grow it. Organic horticulture, or organic gardening, has become increasingly popular for the modern home gardener.

There are many types of vegetable gardens. The potager, a garden in which vegetables, herbs and flowers are grown together, has become more popular than the more traditional rows or blocks.

Herb Garden

The herb garden is often a separate space in the garden, devoted to growing a specific group of plants known as herbs. These gardens may be informal patches of plants, or they may be carefully designed, even to the point of arranging and clipping the plants to form specific patterns, as in a knot garden.

Herb gardens may be purely functional or they may include a blend of functional and ornamental plants. The herbs are usually used to flavour food in cooking, though they may also be used in other ways, such as discouraging pests, providing pleasant scents, or serving medicinal purposes (such as a physic garden), among others.

A kitchen garden can be created by planting different herbs in pots or containers, with the added benefit of mobility. Although not all herbs thrive in pots or containers, some herbs do better than others. Mint is an example of a herb that is advisable to keep in a container or it will take over the whole garden.

Some popular culinary herbs in temperate climates are to a large extent still the same as in the medieval period.

Herbs often have multiple uses. For example, mint may be used for cooking, tea, and pest control. Examples of herbs and their uses (not intended to be complete):

- Annual culinary herbs: basil, dill, summer savory
- Perennial culinary herbs: mint, rosemary, thyme, tarragon
- Herbs used for potpourri: lavender, lemon verbena
- Herbs used for tea: mint, lemon verbena, chamomile, bergamot, hibiscus
- Herbs used for other purposes: stevia for sweetening, feverfew for pest control in the garden.

2

Cole Crops

Cole crops are a common sight in the home garden, especially in cooler weather. But some gardeners may not know what cole crops are. Whether you know what cole crop plants are or not, chances are you enjoy them on a regular basis.

What are Cole Crops?

Cole crops, at a basic level, are plants that belong to the mustard (Brassica) family and are all descendants of wild cabbage. As a group, these plants grow better in cool weather. This leads many people to think that the word "cole" is a variation of the word "cold" and they may even refer to these plants as cold crops. Actually, the word "cole" is a variation of a Latin word that means stem.

When to Plant Cole Crops

The specific time when to plant cole crops will be different depending on which one you are growing. For example, most cabbage varieties can be planted out much earlier than broccoli or cauliflower because cabbage plants can tolerate a much lower temperature. In general, these crops grow best when daytime temperatures are below 80 degrees F. and night-time temperatures are below 60 degrees F. at night. Temperatures higher than this can lead to buttoning, bolting or poor head formation.

Growing Cole Crop Plants

For best results, cole crops should be grown in full sun, but due to their need for cooler temperatures, if you have a partially shaded garden, the vegetables in this family will do okay here as well. Also, if you live in an area that has a short cool season, planting them in part shade can help alleviate daytime temperatures by keeping direct sun from falling on the plants.

Cole crop plants usually need significant amounts of nutrients, particularly micro-nutrients that may not be found in standard fertilizers. Therefore, it is important to work organic material into the beds you plan on growing cole crops in prior to planting them.

Since many of these crops are susceptible to the same types of diseases and pests, rotating plants at least every few years is a good idea. This will help to cut down on diseases and pests that overwinter in the soil and attack the plants.

Brussels Sprout

Figure: Brussels sprout plants

Brussel sprouts, Brassica oleracea, are a cultivar of cabbage in the family Brassicaceae grown for their edible small leafy green buds, which resemble miniature cabbages. The plant has long, smooth and leathery leaves which can be green to purple in colour and are arranged alternately on the stem. The sprouts form at the base of each leaf, in a long, spiral stem. The edible portion of the crop is the bud, which is a small cabbage-like head. It is a light green to green/blue in colour.

The plant is biennial but grown as an annual and can reach 0.6–1 m (2–3 ft) in height with a spread of 0.5–0.6 m (1.5–2 ft). Brussel sprouts may also be simply referred to as 'sprouts' and they originated from Northern Europe.

Uses

Brussel sprouts are most commonly eaten cooked as a vegetable. They may also be pickled in vinegar.

Propagation

Requirements: Brussels sprouts are a cool season crop which can be grown both in Spring and in Fall. The plants thrive in cool climates, maturing in cool or lightly frosty weather. In areas with hot summers, they should be planted for a Fall harvest. Brussels sprouts grow best in moist, fertile, well-draining soil with a slightly acidic pH between 6.0 and 6.5 and at temperatures between 7 and 24°C (45–75°F), with optimum growth occurring at 15–18°C (59–64°F).

Brussels sprouts have a high nitrogen requirement and due to the reduced activity of soil microbes in late fall and winter, organic matter should be added to the soil throughout the year to ensure an adequate supply of nutrients when sprouts are planted. Plant Brussels sprouts in an area that receives at least six hours of full sunlight for optimum growth and development.

Sowing Seeds

Brussels sprouts can be direct seeded or started indoors for transplants. Spring plantings should be made 2–3 weeks before the last frost date in your area and Fall plantings should be made approximately 4 months before the first Fall frost. Sow seed 1.3 cm (0.5 in) deep in small groups of 2–3 seeds and about a week after emergence, thin to a final spacing of 60 cm (24 in) within the row, allowing 76 cm (30 in) between rows. Keep soil evenly moist after planting.

If starting indoors, plant seed in peat pots to minimize disturbance to the roots when transplanting. Seedling can be planted outdoors when they are 3–4 weeks old at the same time as seeds are planted using the spacing detailed above. Plant transplants slightly deeper in the ground than they currently are in their pot and keep soil moist to ensure good fertility.

General Care

Brussels sprouts are sensitive to boron deficiency which can cause the plants to develop hollow stems and small buds.

The deficiency can be corrected by adding boron to the soil. Brussels sprouts should be provided with adequate and even moisture (1.0 to 1.5 inches a week) to keep plants fertile and prevent them from bolting.

Mulching around the plants helps to conserve soil moisture and reduces the soil temperature.

Harvesting

Brussels sprouts are generally ready for harvest 90 to 180 days after planting when the heads are firm and green and have reached between 2.5 and 5.0 cm (1–2 in) in diameter.

The oldest sprouts are located at the bottom of the stalk and mature upwards. As sprouts are removed, it can be beneficial to remove the leaves which are beginning to yellow.

Some people prefer to remove all of the leaves on the stalk to accelerate maturity and harvest in a manner similar to commercial sprout production.

The stalks can be removed before the ground freezes over and stored indoors for further harvesting.

Common Pests and Diseases

Black rot *Xanthomonas campestris*

Symptoms

Seedlings develop wilted yellow to brown leaves and collapse; yellow, V-shaped lesions on mature leaf margins; dark rings can be found in the cross section of the stem

Cause: Bacteria

Comments

Can be confused with Fusarium wilt; favours warm wet environments

Management

Primary control methods based on good sanitation; plant disease-free seed; rotate crops every 2 years or less to non-brassica; avoid sprinkler irrigation

Cabbage aphid *Brevicoryne brassicae*

Figure: Cabbage aphids on Brussels sprouts

Symptoms

Large populations can cause stunted growth or even plant death; insects visible on the plant leaves or sprouts and are small, grey-green in colour and soft bodied and are covered with a white waxy coating; leaves may curl; insect excretes a sticky substance called honedew

Cause: Insect

Comments

Cabbage aphids feed only on cruciferous plants but may survive on related weed species

Management

If aphid population is limited to just a few leaves or shoots then the infestation can be pruned out to provide control; check transplants for aphids before planting; use tolerant varieties if available; reflective mulches such as silver coloured plastic can deter aphids from feeding on plants; sturdy plants can be sprayed with a strong jet of water to knock aphids from leaves; insecticides are generally only required to treat aphids if the infestation is very high - plants generally tolerate low and medium level infestation; insecticidal soaps or oils such as neem or canola oil are usually the best method of control; always check the labels of the products for specific usage guidelines prior to us

Flea Beetle Systena Blanda

Phyllotreta spp.

Figure: *Flea beetles are often metallic in appearance*

Symptoms

Small holes or pits in leaves that give the foliage a characteristic "shothole" appearance; young plants and seedlings are particularly susceptible; plant growth may be reduced; if damage is severe the plant may be killed; the pest responsible for the damage is a small (1.5–3.0 mm) dark coloured beetle which jumps when disturbed; the beetles are often shiny in appearance

Cause: Insect

Comments

Younger plants are more susceptible to flea beetle damage than older ones; older plants can tolerate infestation; flea beetles may overwinter on nearby weed species, in plant debris or in the soil; insects may go through a second or third generation in one year

Management

In areas where flea beetles are a problem, floating row covers may have to be used prior to the emergence of the beetles to provide a physical barrier to protect young plants; plant seeds early to allow establishment before the beetles become a problem - mature plants are less susceptible to damage; trap crops may provide a measure of control - cruciferous plants are best; application of a thick layer of mulch may help prevent beetles reaching surface; application on diamotecoeus earth or oils such as neem oil are effective control methods for organic growers; application of insecticides containing carbaryl, spinosad, bifenthrin and permethrin can provide adequate control of beetles for up to a week but will need reapplied

Downy Mildew: *Hyaloperonospora parasitica*

Symptoms

Small angular lesions on upper surface of leaves which enlarge into orange or yellow necrotic patches; white fluffy growth on undersides of leaves

Cause: Fungus

Comments

Disease emergence favoured by cool, moist conditions

Management

Remove all crop debris after harvest; rotate with non-brassicas; it is possible to control downy mildew on brussel sprouts with the application of an appropriate fungicide

Powdery Mildew: *Erysiphe cruciferarum*

Symptoms

Small white patches on upper and lower leaf surfaces which may also show purple blotching; patches coalesce to form a dense powdery layer which coats the leaves; leaves become chlorotic and drop from plant.

Cause: Fungus

Comments

Disease emergence favoured by dry season, moderate temperatures, low humidity and low levels of rainfall

Management

Plant resistant varieties; rotate crops; remove all crop debris after harvest; remove weeds; avoid excessive application of nitrogen fertilizer which encourages powdery mildew growth; powdery mildew can be controlled by application of sulfur sprays, dusts or vapors

White Rust: *Albugo candida*

Symptoms

White pustules on cotyledons, leaves, stems and/or flowers which coalesce to form large areas of infection; leaves may roll and thicken

Cause: Fungus

Comments

Fungus can survive for long periods of time in dry conditigns; disease spread by wind

Management

Rotate crops; plant only disease-free seed; apply appropriate fungicide if disease becomes a problem

Fusarium wilt (Yellows) *Fusarium oxysporum*

Symptoms

Dull yellow-green discolouration of leaves and stunted plant growth; lower leaves turning yellow-brown necrotic areas and dropping from plant; browning and defoliation spreading upwards from bottom of plant; symptoms more pronounced on one side of leaf than the other

Cause: Fungus

Comments

Fungus can survive for many years in soil and can enter plant through root tips or wounds; disease generally favours warm soil temperatures

Management

Plant resistant varieties; most other control methods are ineffective once disease has emerged; spread can be prevented by disinfecting equipment

Ring spot: *Mycosphaerella brassicicola*

Symptoms

Small, purple spots surrounded by a ring of water-soaked tissue on leaves which mature to brown spots with olive green borders 1-2 cm across; spots may develop numerous fruiting bodies which give them a black appearance or develop a concentric pattern; heavily infected leaves may dry up and curl inwards

Cause: Fungus

Comments

Ring spot requires cool, moist conditions to survive; disease symptoms typically develop in the fall and the peak of the infection occurs in winter

Management

Refrain from planting in areas known to have had disease previously; rotate crop to non-brassicas; sanitize tools and equipment regularly; apply appropriate fungicide if disease is identified in crop

Alternaria leaf spot (Black spot, Gray spot) *Alternaria brassicae*

Symptoms

Small dark spots on leaves which turn brown to gray; lesions may be round or angular and may possess a purple-black margin; lesions may form concentric rings, become brittle and crack in centre; dark brown elongated lesions may develop on stems and petioles

Cause: Fungus

Comments

May become a problem during cool, wet periods

Management

Plant only pathogen-free seed; rotate crops; applications of appropriate fungicides control disease when present

Clubroot: *Plasmodiophora brassicae*

Symptoms

Slow growing, stunted plants; yellowish leaves which wilt during day and rejuvenate in part at night; swollen, distorted roots; extensive gall formation

Cause: Fungus

Comments

Can be difficult to distinguish from nematode damage; fungus can survive in soil for periods in excess of 10 years; can be spread by movement of contaminated soil and irrigation water to uninfected areas

Management

Once the pathogen is present in the soil it can survive for many years, elimination of the pathogen is economically unfeasible; rotating crops generally does not provide effective control; plant only certified seed and avoid field grown transplants unless produced in a fumigated bed; applying lime to the soil can reduce fungus sporulation

Sclerotinia stem rot (White mold) *Sclerotinia sclerotiorum*

Symptoms

Irregular, necrotic gray lesions on leaves; white-gray leions on stems; reduced pod set; shattering seed pods

Cause: Fungus

Comments

Disease emergence favours moderate to cool temperatures and high humidity

Management

Rotate crop to non-hosts (e.g. cereals) for at least 3 years; control weeds; avoid dense growth by planting in adequately spaced rows; apply appropriate foliar fungicides

Large cabbage white (Cabbageworm) *Pieres rapae*

Figure: Cabbageworm and frass

Symptoms

Large ragged holes in leaves or bored into sprouts; green-brown frass (insect feces) on leaves; caterpillar is green in colour and hairy, with a velvet-like appearance; may have faint yellow to orange stripes down back; slow-moving compared with other caterpillars

Cause: Insect

Comments

Butterfly larvae cause damage by feeding on plants; can be distinguished from other caterpillars by its sluggish movement; in large numbers larvae can cause extensive damage very quickly

Management

Hand pick caterpillars from plants and destroy; scrape eggs from leaves prior to hatching; apply appropriate insecticide if infestation is very heavy

Diamondback Moth: *Plutella xylostella*

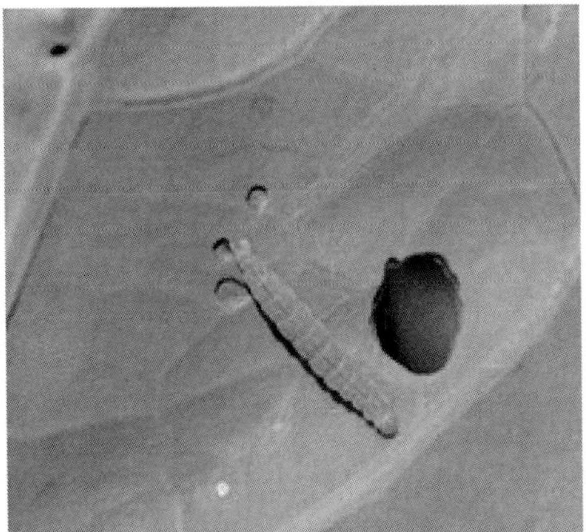

Figure: *Diamondback moth larva*

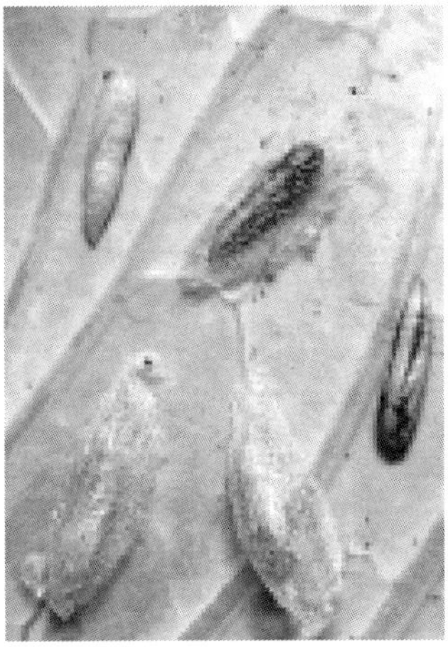

Figure: Diamondback moth pupae

Symptoms

Young larvae feed between upper and lower leaf surface and may be visible when they emerge from small holes on the underside of the leaf; older larvae leave large, irregularly shaped shotholes on leaf undersides, leaving the upper surface intact; larvae may drop from the plant on silk threads if the leaf is disturbed; larvae are small (1 cm/ 0.3 in) and tapered at both ends; larvae have to prolegs at the rear end that are arranged in a distinctive V-shape

Cause: Insect

Comments

Larvae take between 10 and 14 days to mature and spin a loose, gauze-like cocoon on leaves or stems to pupate

Management

Larvae can be controlled organically by applications of *Bacillus thurengiensis* or Entrust; application of appropriate chemical insecticide is only necessary if larvae are damaging the growing tips of the plants

Cabbage Looper: *Trichoplusia ni*

Figure: Cabbage looper and damage

Symptoms

Large or small holes in leaves; damage often extensive; caterpillars are pale green with a white lines running down either side of their body; caterpillars are easily distinguished by the way they arch their body when moving; eggs are laid singly, usually on the lower leaf surface close to the leaf margin, and are white or pale green in colour

Cause: Insect

Comments

Insects overwinter as pupae in crop debris in soil; adult insect id a dark coloured moth; caterpillars have a wide host range

Management

Looper populations are usually held in check by natural enemies; if they do become problematic larvae can be hand-picked from the plants; an organically acceptable control method is the application of *Bacillus thuringiensis* which effectively kills younger larvae; chemical sprays may damage populations of natural enemies and should be selected carefully

Thrips (Western Flower Thrips, Onion Thrips, etc): *Frankliniella occidentalis*

Thrips tabaci

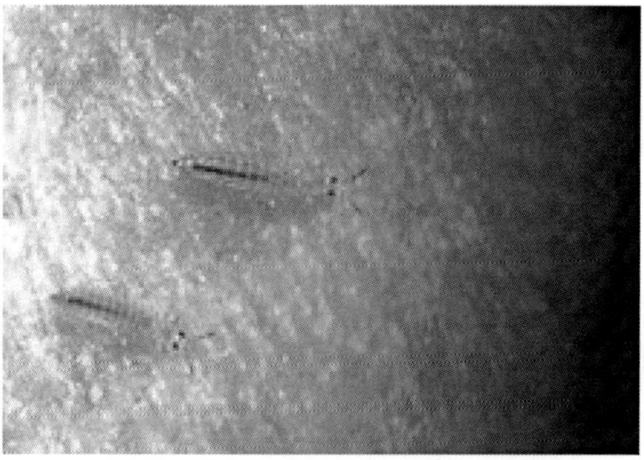

Figure: Western flower thrips

Symptoms

If population is high leaves may be distorted; leaves are covered in coarse stippling and may appear silvery; leaves speckled with black feces; insect is small (1.5 mm) and slender and best viewed using a hand lens; adult thrips are pale yellow to light brown and the nymphs are smaller and lighter in colour

Cause: Insect

Transmit viruses such as Tomato spotted wilt virus; once acquired, the insect retains the ability to transmit the virus for the remainder of its life

Management

Avoid planting next to onions, garlic or cereals where very large numbers of thrips can build up; use reflective mulches early in growing season to deter thrips; apply appropriate insecticide if thrips become problematic

Root Knot Nematode: *Meloidogyne* spp.

Symptoms

Galls on roots which can be up to 3.3 cm (1 in) in diameter but are usually smaller; reduction in plant vigor; yellowing plants which wilt in hot weather

Cause: Nematode

Comments

Galls can appear as quickly as a month prior to planting; nematodes prefer sandy soils and damage in areas of field or garden with this type of soil is most likely

Management

Plant resistant varieties if nematodes are known to be present in the soil; check roots of plants mid-season or sooner if symptoms indicate nematodes; solarizing soil can reduce nematode populations in the soil and levels of inoculum of many other pathogens

Cabbage

Cabbage (*Brassica oleracea* or variants) is a leafy green or purple biennial plant, grown as an annual vegetable crop for its dense-leaved heads. Closely related to other cole crops, such as broccoli, cauliflower, and brussels sprouts, it descends from *B. oleracea var. oleracea*, a wild field cabbage. Cabbage heads generally range from 0.5 to 4 kilograms (1 to 9 lb), and can be green, purple and white. Smooth-leafed firm-headed green cabbages are the most common, with smooth-leafed red and crinkle-leafed savoy cabbages of both colours seen more rarely. It is a multi-layered vegetable. Under conditions of long sunlit days such as are found at high northern latitudes in summer, cabbages can grow much larger. Some records are discussed at the end of the history section.

It is difficult to trace the exact history of cabbage, but it was most likely domesticated somewhere in Europe before 1000 BC, although savoys were not developed until the 16th century. By the Middle Ages, it had become a prominent part of European cuisine. Cabbage heads are generally picked during the first year of the plants' life cycles, but those intended for seed are allowed to grow a second year, and must be kept separated from other cole crops to prevent cross-pollination. Cabbage is prone to several nutrient deficiencies, as well as multiple pests, bacteria and fungal diseases.

The Food and Agriculture Organisation of the United Nations (FAO) reports that world production of cabbage and other brassicas for 2011 was almost 69 million metric tons (68 million long tons; 75 million short tons). Almost half of these crops were grown in China, where Chinese cabbage is the most popular *Brassica* vegetable. Cabbages are prepared in many different ways for eating. They can be pickled, fermented for dishes such as sauerkraut, steamed, stewed, sautéed, braised, or eaten raw. Cabbage is a good source of vitamin K, vitamin C and dietary fibre. Contaminated cabbage has been linked to cases of food-borne illness in humans.

Cabbage (*Brassica oleracea* or *B. oleracea* var. *capitata*, var. *tuba*, var. *sabauda* or var. *acephala*) is a member of the genus *Brassica* and the mustard family, Brassicaceae. Several other cruciferous vegetables (sometimes known as *cole crops*) are considered cultivars of *B. oleracea*, including broccoli, collard greens, brussels sprouts, kohlrabi and sprouting broccoli. All of these developed from the wild cabbage *B. oleracea* var. *oleracea*, also called colewort or field cabbage. This original species evolved over thousands of years into those seen today, as selection resulted in cultivars having different characteristics, such as large heads for cabbage, large leaves for kale and thick stems with flower buds for broccoli. The varietal epithet *capitata* is derived from the Latin word for "having a head". *B. oleracea* and its derivatives have hundreds of common names throughout the world.

Description

Cabbage seedlings have a thin taproot and cordate (heart-shaped) cotyledons. The first leaves produced are ovate (egg-shaped) with a lobed petiole. Plants are 40–60 cm (16–24 in) tall in their first year at the mature vegetative stage, and 1.5–2.0 m (4.9–6.6 ft) tall when flowering in the second year. Heads average between 1 and 8 pounds (0.5 and 4 kg), with fast-growing, earlier-maturing varieties producing smaller heads. Most cabbages have thick, alternating leaves, with margins that range from wavy or lobed to highly dissected; some varieties have a waxy bloom on the leaves. Plants have root systems that are fibrous and shallow. About 90 percent of the root mass is in the upper 20–30 cm (8–12 in) of soil; some lateral roots can penetrate up to 2 m (6.6 ft) deep.

The inflorescence is an unbranched and indeterminate terminal raceme measuring 50–100 cm (20–40 in) tall, with flowers that are yellow or white. Each flower has four petals set in a perpendicular pattern, as well as four sepals, six stamens, and a superior ovary that is two-celled and contains a single stigma and style. Two of the six stamens have shorter filaments. The fruit is a silique that opens at maturity through dehiscence to reveal brown or black seeds that are small and round in shape. Self-pollination is impossible, and plants are cross-pollinated by insects. The initial leaves form a rosette shape comprising 7 to 15 leaves, each measuring 25–35 cm (10–14 in) by 20–30 cm (8–12 in); after this, leaves with shorter petioles develop and heads form through the leaves cupping inward.

Many shapes, colours and leaf textures are found in various cultivated varieties of cabbage. Leaf types are generally divided between crinkled-leaf, loose-head savoys and smooth-leaf firm-head cabbages, while the colour spectrum includes white and a range of greens and purples. Oblate, round and pointed shapes are found.

Cabbage has been selectively bred for head weight and morphological characteristics, frost hardiness, fast growth and storage ability. The appearance of the cabbage head has been given importance in selective breeding, with varieties being chosen for shape, colour, firmness and other physical characteristics. Breeding objectives are now focused on increasing resistance to various insects and diseases and improving the nutritional content of cabbage. Scientific research into the genetic modification of B. oleracea *crops, including cabbage, has included European Union and United States explorations of greater insect and herbicide resistance. Genetically modified* B. oleracea *crops are not currently used in commercial agriculture.*

Cultivation

Cabbage is generally grown for its densely leaved heads, produced during the first year of its biennial cycle. Plants perform best when grown in well-drained soil in a location that receives full sun. Different varieties prefer different soil types, ranging from lighter sand to heavier clay, but all prefer fertile ground with a pH between 6.0 and 6.8. For optimal growth, there must be adequate levels of nitrogen in the soil, especially during the early head formation stage, and sufficient phosphorus and potassium during the early stages of expansion of the outer leaves. Temperatures between 4 and 24 °C (39 and 75 °F) prompt the best growth, and extended periods of higher or lower temperatures may result in premature bolting (flowering). Flowering induced by periods of low temperatures (a process called vernalization) only occurs if the plant is past the juvenile period.

The transition from a juvenile to adult state happens when the stem diameter is about 6 mm (0.24 in). Vernalization allows the plant to grow to an adequate size before flowering. In certain climates, cabbage can be planted at the beginning of the cold period and survive until a later warm period without being induced to flower, a practice that was common in the eastern US.

Plants are generally started in protected locations early in the growing season before being transplanted outside, although some are seeded directly into the ground from which they will be harvested. Seedlings typically emerge in about 4–6 days from seeds planted 1.3 cm (0.5 in) deep at a soil temperature between 20 and 30 °C (68 and 86 °F). Growers normally place plants 30 to 61 cm (12 to 24 in) apart. Closer spacing reduces the resources available to each plant (especially the amount of light) and increases the time taken to reach maturity. Some varieties of cabbage have been developed for ornamental use; these are generally called "flowering cabbage".

They do not produce heads and feature purple or green outer leaves surrounding an inner grouping of smaller leaves in white, red, or pink. Early varieties of cabbage take about 70 days from planting to reach maturity, while late varieties take about 120 days. Cabbages are mature when they are firm and solid to the touch. They are harvested by cutting the stalk just below the bottom leaves with a blade.

The outer leaves are trimmed, and any diseased, damaged, or necrotic leaves are removed. Delays in harvest can result in the head splitting as a result of expansion of the inner leaves and continued stem growth. Factors that contribute to reduced head weight include: growth in the compacted soils that result from no-till farming practices, drought, waterlogging, insect and disease incidence, and shading and nutrient stress caused by weeds.

When being grown for seed, cabbages must be isolated from other B. oleracea *subspecies, including the wild varieties, by 0.8 to 1.6 km (0.5 to 1 mi) to prevent cross-pollination. Other* Brassica *species, such as* B. rapa, B. juncea, B. nigra, B. napus *and* Raphanus sativus, *do not readily cross-pollinate.*

Cultivars

Figure: *White cabbage*

There are several cultivar groups of cabbage, each including many cultivars:

- Savoy – Characterized by crimped or curly leaves, mild flavour and tender texture
- Spring Greens – Loose-headed, commonly sliced and steamed
- Green – Light to dark green, slightly pointed heads. This is the most commonly grown cultivar.
- Red – Smooth red leaves, often used for pickling or stewing
- White, also called Dutch – Smooth, pale green leaves

Some sources only delineate three cultivars: savoy, red and white, with spring greens and green cabbage being subsumed into the latter.

Cultivation Problems

Due to its high level of nutrient requirements, cabbage is prone to nutrient deficiencies, including boron, calcium, phosphorus and potassium. There are several physiological disorders that can affect the postharvest appearance of cabbage. Internal tip burn occurs when the margins of inside leaves turn brown, but the outer leaves look normal. Necrotic spot is where there are oval sunken spots a few millimetres across that are often grouped around the midrib. In pepper spot, tiny black spots occur on the areas between the veins, which can increase during storage.

Fungal diseases include wirestem, which causes weak or dying transplants; *Fusarium* yellows, which result in stunted and twisted plants with yellow leaves; and blackleg, which leads to sunken areas on stems and gray-brown spotted leaves. The fungi *Alternaria brassicae* and *A. brassicicola* cause dark leaf spots in affected plants. They are both seedborne and airborne, and typically propagate from spores in infected plant debris left on the soil surface for up to twelve weeks after harvest. *Rhizoctonia solani* causes the post-emergence disease wirestem, resulting in killed seedlings ("damping-off"), root rot or stunted growth and smaller heads.

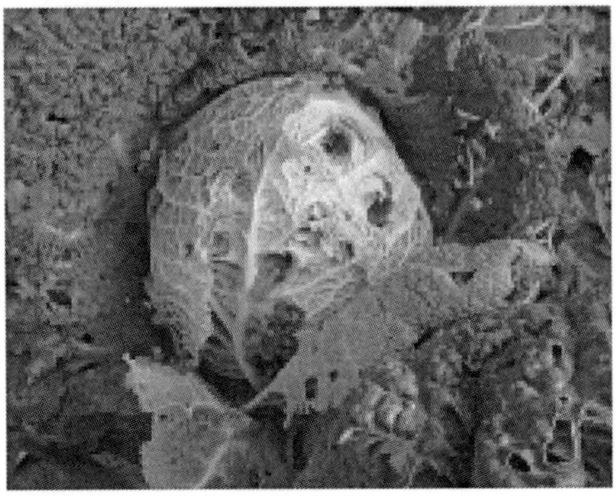

Figure: *Cabbage moth damage to a savoy*

One of the most common bacterial diseases to affect cabbage is black rot, caused by *Xanthomonas campestris*, which causes chlorotic and necrotic lesions that start at the leaf margins, and wilting of plants. Clubroot, caused by the soilborne slime mold-like organism *Plasmodiophora brassicae*, results in swollen, club-like roots. Downy mildew, a parasitic disease caused by the oomycete *Peronospora parasitica*, produces pale leaves with white, brownish or olive mildew on the lower leaf surfaces; this is often confused with the fungal disease powdery mildew.

Pests include root-knot nematodes and cabbage maggots, which produce stunted and wilted plants with yellow leaves; aphids, which induce stunted plants with curled and yellow leaves; harlequin bugs, which cause white and yellow leaves; thrips, which lead to leaves with white-bronze spots; striped flea beetles, which riddle leaves with small holes; and caterpillars, which leave behind large, ragged holes in leaves. The caterpillar stage of the "small cabbage white butterfly" (*Pieris rapae*), commonly known in the United States as the "imported cabbage worm", is a major cabbage pest in most countries. The large white butterfly (*Pieris brassicae*) is prevalent in eastern European countries. The diamondback moth (*Plutella xylostella*) and the cabbage moth

(*Mamestra brassicae*) thrive in the higher summer temperatures of continental Europe, where they cause considerable damage to cabbage crops. In India, the diamondback moth has caused losses up to 90 percent in crops that were not treated with insecticide. Destructive soil insects include the cabbage root fly (*Delia radicum*) and the cabbage maggot (*Hylemya brassicae*), whose larvae can burrow into the part of plant consumed by humans.

Planting near other members of the cabbage family, or where these plants have been placed in previous years, can prompt the spread of pests and disease. Excessive water and excessive heat can also cause cultivation problems.

The Pests

There are many pests in the vegetable garden which can completely destroy a cabbage crop. The commonly found cabbage pests listed here are also found on other plants in the garden.

- Aphids appear in groups or colonies and are small black, brown, gray, green, pinkish or yellow globular-shaped insects most often found on the underside of leaves. Treat infestations with insecticidal soap spray or a commercial spray specifically formulated to treat aphids. Spray foliage with water periodically to disperse the insects.
- Cabbage Loopers are green caterpillars that should be hand-picked from leaves as soon as they appear. At the least, caterpillars will eat holes through leaves. If not removed, they will tunnel through leaves in a matter of days and may eat entire leaf so there is nothing left but cut stems. Using row covers prevents pests from attacking young plants. Commercial insecticidal dust and spray is also available to control caterpillars.
- Cutworms are gray caterpillars in or on top of the soil under young plants. Cutworms eat plants at ground level. Keep garden healthy by weeding regularly. Cultivate soil around plants, pick out and destroy any caterpillars found. Commercial insecticide may also be used.
- Flea Beetles are easy to spot on a sunny day because the sun reflects off their shiny bodies. They are black, some with lighter-coloured stripes and can be seen hopping from plant to plant. Flea beetles are most often seen in the garden during hot, dry periods. Provide shade when temperatures are extreme and keep garden well watered to help keep population down. Commercial insecticide may also be used.

- Onion Thrips are small, elongated brownish bugs with lighter stripes on each side. These insects may cause leaves to shrivel, appear whitish to silver in colour and be riddled with tiny holes giving foliage the appearance of fine lace. Treat the same as with flea beetle infestation.
- Rodents such as gophers, moles, rabbits, voles and others will eat young plants or tunnel through garden soil damaging the root systems which may provide an opening for disease. The best way to keep these pests out of the garden is by surrounding it with a wire fence that has been partially buried.
- Whiteflies are small flies with white wings. They are found on the undersides of leaves. When disturbed, whiteflies fill the air around plants. This pest is found in greenhouses and outdoor gardens alike. The heat and humidity in the southern states provide ideal conditions for infestation. Treat infested plants with a natural insecticidal soap solution early in the day. Retreat as needed.

Cabbage is not a hard vegetable to grow. It may take a little extra attention than some others but it is worth the effort. Remember to begin with healthy soil and be certain to plant cabbage in rotation to eliminate the spread of disease found in soil where these plants are grown. Intersperse companion plants to invite natural enemies such as ladybugs and parasitic wasps to the garden.

Cauliflower

Cauliflower is one of several vegetables in the species *Brassica oleracea*, in the family Brassicaceae. It is an annual plant that reproduces by seed. Typically, only the head (the *white curd*) is eaten. The cauliflower head is composed of a white inflorescence meristem. Cauliflower heads resemble those in broccoli, which differs in having flower buds.

Its name is from Latin *caulis* (cabbage) and flower,. *Brassica oleracea* also includes broccoli, brussels sprouts, cabbage, collard greens, and kale, though they are of different cultivar groups.

Insect Pests of Cauliflower and Their Management

Cauliflower (Brassica oleracea L. var. botrytis) is an important winter vegetable grown in India. It belongs to the family Cruciferae. India produces about 4.694 M mt of Cauliflower per year from 0.256 M ha area with an average productivity of 18.3 mt/ha. Bihar, Uttar Pradesh, West Bengal, Orissa, Assam, Haryana, Rajasthan and Maharashtra are major cauliflower growing states. It also cultivated in non traditional areas of Tamil Nadu, Andhra Pradesh and Kerala. It is consumed as vegetable in curries, soups and pickles and low in fat and rich source of dietary fibres. In India, the major constraints for production loss are due to physiological disorders, pests and diseases. Off these, insect pests are prime important as it cause serious economic damage to this crop.

The pests such as Diamond Back Moth (DBM), Plutella xylostella, Tobacco caterpillar Spodoptera litura, Leaf webber Crocidolomia binotalis, stem borer Hellula undalis, Aphid, Brevicornea brassicae, Mustard aphid, Lipaphis erisimi and Painted bugs, Bagrada cruciferarum are the important pests of cauliflower. Among these, DBM and tobacco caterpillar are notorious and chemical insecticides are widely used by the farmers to control these pests. Sole relay on synthetic chemicals is not recommended while considering pesticide residues as well other issues like insecticide resistance, resurgence and secondary pests out break. Keeping the facts in view, here we discussed some the integrated approaches for the sustainable pest management programme and it would be highly useful for cauliflower growers across the country.

Basic Information

Problem: Cabbage worms, Cabbage loopers

Affected Area: Leaf

Description: Loopers are caterpillars that are generally green or greyish, taking on the colour of the host plant. They may blend well and be difficult to see. They eventually develop into white- or yellow-winged butterflies often seen fluttering about the plants. Caterpillars feed on the underside of leaves leaving ragged holes sometimes to the extent that plants starve and die.

Control: Spray with Bacillus thuringiensis (Bt) Comercially available as Dipel or Thuricide.) Cover with remay or similiar product in spring.

Problem: Cutworms

Affected Area: Stem

Description: Plants chewed off just above ground level. Cutworms are caterpillars 1 1/2" long and mottled or striped green, brown or gray. When they are disturbed, they roll up in a coil. They usually position themselves at the moisture line in the soil moving up and down according to the water content. If the soil surface is dry, they will be found a couple of inches below the surface where the moisture begins. When newly watered, they will be at the surface.

Control: Put cardboard collar around new transplants to extend 1" to 2" above and below soil level.

Problem: Root Maggots

Affected Area: Entire Plant

Description: Leaves wilt and growth is stunted. Insect is common, white, root feeding maggot. About the time the cherries bloom, 1/4 inch long adult flies emerge from the soil and lay their eggs at base of plants in surrounding soil. The legless larvae feed on host plants for three weeks, riddling the roots with brown tunnels before they pupate. Two or three generations can occur each growing season.

Control: Use Diazinon before planting as a soil treatment. Rotate from year to year.

Problem: Flea Beetle

Affected Area: Leaf and Root

Description: Tiny holes ?pinholes? chewed in leaves by adult insect. Adult insects are 1/16" long, hard shelled, shiny, dark-coloured beetles that jump when disturbed. Slender, whitish, cylindrical larvae feed in or on roots but root damage is generally minimal.

Control: Dust with Rotenone. Keep debris removed. Rotate location of planting from year to year.

Problem: Aphids

Affected Area: Leaf

Description: Green, red black or white insects that cause curled yellow leaves and exude a honeydew substance

Control: Insecticidal soaps or a strong stream of water or most labelled insecticides like Diazinon or Sevin. A layer of aluminum foil under plants reflects light to underside of leaves making them an undesirable habitat for aphids.

Problem: Slugs and Snails

Affected Area: Entire Plant

Description: Large portions of young plants missing.

Control: Slugs and snails are very susceptible to desiccation (drying) and require a moist, shady place to live. Cultural practices which promote a sunny, dry environment will discourage them. Avoid too-frequent waterings allowing soil surface to dry out between irrigations. Keep garden free of debris, boards, bricks, and stones where they hide. Hand picking these pests is very effective. Create ?traps? for hand picking by laying boards in the garden. Slugs and snails will congregate under them. Lift the boards each morning and collect the slugs and snails. Dispose of them completely as they will crawl back if tossed out of the garden and eggs inside dead pests can still hatch to produce more of these pests. Slug and snail bait containing metaldehyde can be placed near food plants as long as they do not contact edible portions of the crop. Most effective when moistened, but not water logged.

Snail bait attracts slugs and snails from several feet away so bait stations are effective. Stations help protect birds, pets and other non-target animals which are also attracted to the bait. Place small piles of bit under a slightly propped up board or use container such as a cottage cheese or yogurt carton. Bury carton to the mouth of the container. Place small amount of commercial bait inside and moisten with apple juice, orange juice or water. Cut hole in lid to allow access and place lid on container.

Containers may also hold beer or yeast water to attract slugs and snails in where they drown. Place bait stations wherever slugs and snails are active or around perimeter of garden.

Problem: Damping Off

Affected Area: Seedling

Description: Young seedlings wilt and die

Control: Use treated seed and let soil dry out between waterings

Problem: Bacterial soft rot

Affected Area: Leaf

Description: Leaves turn yellow (chlorotic) beginning at margins and spreading inwards. Veins within area turn black. Infection enters main stem turning the inside black.. Plants either die or are dwarfed when young, become defoliated if more mature.

Control: Plant resistant varieties and rotate crops from year to year.

Collards

Collard greens is the American English term for various loose-leafed cultivars of *Brassica oleracea*, part of the Acephala group which also contains cabbage and broccoli. The plants are grown for their large, dark-coloured, edible leaves and as a garden ornamental, mainly in Brazil, Portugal, the southern United States, many parts of Africa, Montenegro, Bosnia and Herzegovina, southern Croatia, northern Spain and in India. They are classified in the same cultivar group as kale and spring greens, to which they are genetically similar. The name "collard" is a corrupted form of the word "colewort" (the wild cabbage plant).

The plant is also called *couve* in Brazil and in Portugal, *couve galega* or *couve portuguesa* (among several other names) in Cape Verde, *berza* in Spanish-speaking countries, *raštika* in Bosnia and Herzegovina and Croatia and *raštan* in Montenegro and Serbia. In Kashmir, it is called *haak*. In Kenya it is more commonly known by its Swahili name, *sukuma wiki*, and is often confused with kales. In New Zealand, it is called Dalmatian Cabbage.

Description

Figure: *Young collard plants growing in a container*

The cultivar group name Acephala ("without a head" in Greek) refers to the fact that this variety of *B. oleracea* does not have the usual close-knit core of leaves (a "head") like cabbage. The plant is a biennial where winter frost occurs, and perennial in even colder regions. It is also moderately sensitive to salinity. It has an upright stalk, often growing up to two feet tall. The plant is very similar to kale. Popular cultivars of collard greens include 'Georgia Southern', 'Morris Heading', 'Butter Collard' (or *couve manteiga*), and *couve tronchuda.*

Cultivation and Storage

The plant is commercially cultivated for its thick, slightly bitter, edible leaves. They are available year-round, but are tastier and more nutritious in the cold months, after the first frost. For best texture, the leaves should be picked before they reach their maximum size, at which stage the leaves will be thicker and should be cooked differently from the new leaves. Age will not affect flavour. Flavour and texture also depend on the cultivar; the couve manteiga and couve tronchuda are especially appreciated in Brazil and Portugal.

Fresh collard leaves can be stored for up to 10 days if refrigerated to just above freezing (1°C) at high humidity (>95%).

In domestic refrigerators, fresh collard leaves can be stored for about three days. Once cooked, they can be frozen and stored for greater lengths of time.

Common Pests and Diseases

Alternaria Leaf Spot: *Alternaria* spp.

Symptoms

Brown to tan concentric rings with yellow edges on leaves; centres of lesions developing gray to brown soft fungal mold; brown to black lesions with a black border on roots

Cause: Fungus

Comments

Disease emergence favours warm, wet conditions

Management

Plant only pathogen-free seed; rotate crops; applications of appropriate fungicides control disease when present

Anthracnose: *Colletotrichum higginsianum*

Symptoms

Small circular or irregularly shaped dry spots which are gray to straw in colour on leaves; a high number of spots may cause the leaf to die; lesions may coalesce to form large necrotic patches causing leaves to turn yellow and wilt; lesions may split or crack in dry centres; dry sunken spots on roots which enlarge and turn gray or brown

Cause: Fungus

Comments

Fungus overwinters on leaf debris and on related weeds; disease emergence is favoured by moist, warm conditions

Management

Control of disease depends on sanitary practices; treat seeds with hot water prior to planting; rotate crops; plant in an area with good soil drainage; remove all cruciferous weeds which may act as a reservoir for the fungus

Black Rot: *Xanthomonas campestris*

Symptoms

Seedlings develop wilted yellow to brown leaves and collapse; yellow, V-shaped lesions on mature leaf margins; dark rings can be found in the cross section of the stem

Cause: Bacterium

Comments

Disease emergence favours warm, wet conditions

Management

Primary control methods based on good sanitation; plant disease-free seed; rotate crops every 2 years or less to non-brassica; avoid sprinkler irrigation

Cercospora leaf spot (Frogeye leaf spot) *Cercospora brassicicola*

Symptoms

Angular or circular green to gray spots with brown borders on leaves; plant defoliation may occur in the case of a severe infestation

Cause: Fungus

Comments

Disease emergence favours cool temperatures and wet weather

Management

Plant only certified disease-free seed; avoid overhead irrigation; rotate crops to non-brassica species for 2-3 years; apply appropriate fungicide if disease emerges

Clubroot: *Plasmodiophora brassicae*

Symptoms

Slow growing, stunted plants; yellowish leaves which wilt during day and rejuvenate in part at night; swollen, distorted roots; extensive gall formation

Cause: Fungus

Comments

Can be difficult to distinguish from nematode damage; fungus can survive in soil for periods in excess of 10 years; can be spread by movement of contaminated soil and irrigation water to uninfected areas

Management

Once the pathogen is present in the soil it can survive for many years, elimination of the pathogen is economically unfeasible; rotating crops generally does not provide effective control; plant only certified seed and avoid field grown transplants unless produced in a fumigated bed; applying lime to the soil can reduce fungus sporulation

Downy Mildew: *Peronospora parasitica*

Symptoms

Gray to white fluffy mold on the underside of leaves; tan to yellow dry spots on upper surface of leaves; leaves dropping

Cause: Fungus

Comments

Disease emergence favours cool temperatures; disease spreads quickly in wet conditions

Management

Remove all crop debris after harvest; rotate with non-brassicas; application of appropriate fungicides may be required if symptoms of disease are present

White Leaf Spot: *Pseudocercosporella capsellae*

Symptoms

Small, necrotic, brown spots on leaf tips or margins that matures to light gray or white with the original dark spot in centre; margins of lesions may be darker; lesions may coalesce to form large chlorotic areas and cause defoliation

Cause: Fungus

Comments

Disease emergence favours cool and wet conditions

Management

No known plant resistance to white leaf spot so control relies on cultural practices such as rotating crops and removing weeds; application of appropriate fungicide may help control the disease

Wirestem (Damping-Off): *Rhizoctonia solani*

Symptoms

Death of seedlings after germination; brown-red or black rot girdling stem; seedling may remain upright but stem is constricted and twisted (wirestem)

Cause: Fungus

Comments

Disease emergence favours slow growing, deeply seeded plants and cool, wet soils

Management

Plant pathogen-free seed or transplants that have been produced in sterilized soil; apply fungicide to seed to kill off any fungi; shallow plant seeds or delay planting until soil warms

Beet Armyworm: *Spodoptera exigua*

Symptoms

Singular, or closely grouped circular to irregularly shaped holes in foliage; heavy feeding by young larvae leads to skeletonized leaves; shallow, dry wounds on fruit; egg clusters of 50-150 eggs may be present on the leaves; egg clusters are covered in a whitish scale which gives the cluster a cottony or fuzzy appearance; young larvae are pale green to yellow in colour while older larvae are generally darker green with a dark and light line running along the side of their body and a pink or yellow underside

Cause: Insect

Comments

Insect can go through 3–5 generations a year

Management

Organic methods of controlling the beet armyworm include biological control by natural enemies which parasitize the larvae and the application of *Bacillus thuringiensis*; there are chemicals available for commercial control but many that are available for the home garden do not provide adequate control of the larvae

Cutworms: *Agrotis* spp.

Peridroma saucia

Nephelodes minians and others

Symptoms

Stems of young transplants or seedlings may be severed at soil line; if infection occurs later, irregular holes are eaten into the surface of fruits; larvae causing the damage are usually active at night and hide during the day in the soil at the base of the plants or in plant debris of toppled plant; larvae are 2.5–5.0 cm (1–2 in) in length; larvae may exhibit a variety of patterns and colouration but will usually curl up into a C-shape when disturbed

Cause: Insects

Comments

Cutworms have a wide host range and attack vegetables including asparagus, bean, cabbage and other crucifers, carrot, celery, corn, lettuce, pea, pepper, potato and tomato

Management

Remove all plant residue from soil after harvest or at least two weeks before planting, this is especially important if the previous crop was another host such as alfalfa, beans or a leguminous cover crop; plastic or foil collars fitted around plant stems to cover the bottom 3 inches above the soil line and extending a couple of inches into the soil can prevent larvae severing plants; hand-pick larvae after dark; spread diatomaceous earth around the base of the plants (this creates a sharp barrier that will cut the insects if they try and crawl over it); apply appropriate insecticides to infested areas of garden or field if not growing organically

Flea Beetles: *Phylotreta* spp.

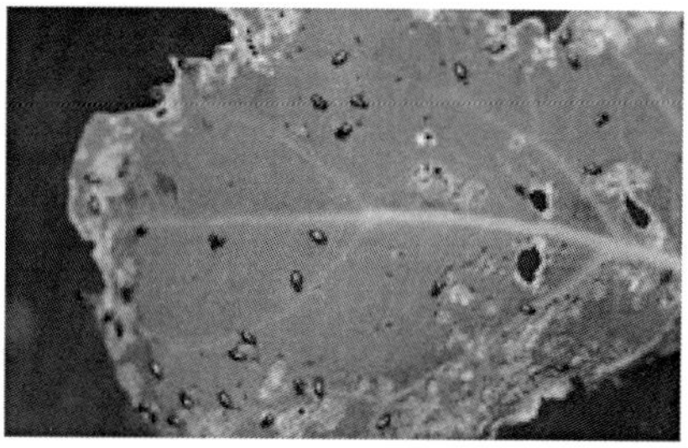

Figure: *Crucifer flea beetle damage on broccoli leaf*

Symptoms

Small holes or pits in leaves that give the foliage a characteristic "shothole" appearance; young plants and seedlings are particularly susceptible; plant growth may be reduced; if damage is severe the plant may be killed; the pest responsible for the damage is a small (1.5–3.0 mm) dark coloured beetle which jumps when disturbed; the beetles are often shiny in appearance

Cause: Insects

Comments

Younger plants are more susceptible to flea beetle damage than older ones; older plants can tolerate infestation; flea beetles may overwinter on nearby weed species, in plant debris or in the soil; insects may go through a second or third generation in one year

Management

In areas where flea beetles are a problem, floating row covers may have to be used prior to the emergence of the beetles to provide a physical barrier to protect young plants; plant seeds early to allow establishment before the beetles become a problem - mature plants are less susceptible to damage; trap crops may provide a measure of control - cruciferous plants are best; application of a thick layer of mulch may help prevent beetles reaching surface; application on diamotecoeus earth or oils such as neem oil are effective control methods for organic growers; application of insecticides containing carbaryl, spinosad, bifenthrin and permethrin can provide adequate control of beetles for up to a week but will need reapplied

Thrips (Western Flower Thrips, Onion Thrips, etc.)
Frankliniella occidentalis

Thrips tabaci

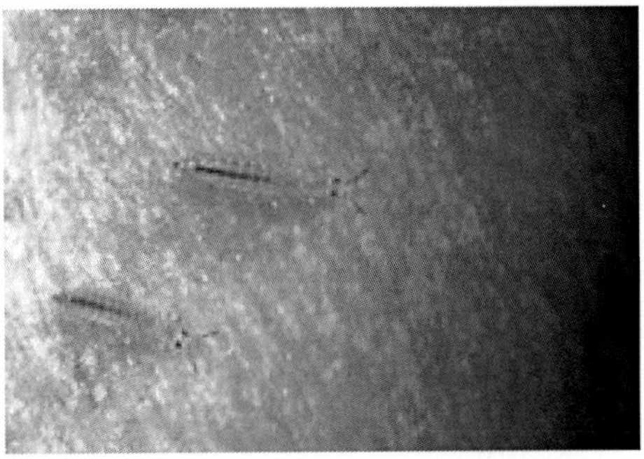

Figure: Western flower thrips

Symptoms

If population is high leaves may be distorted; leaves are covered in coarse stippling and may appear silvery; leaves speckled with black feces; insect is small (1.5 mm) and slender and best viewed using a hand lens; adult thrips are pale yellow to light brown and the nymphs are smaller and lighter in colour

Cause: Insect

Comments

Transmit viruses such as Tomato spotted wilt virus; once acquired, the insect retains the ability to transmit the virus for the remainder of its life

Management

Avoid planting next to onions, garlic or cereals where very large numbers of thrips can build up; use reflective mulches early in growing season to deter thrips; apply appropriate insecticide if thrips become problematic

Root Knot Nematode: *Meloidogyne* spp.

Symptoms

Galls on roots which can be up to 3.3 cm (1 in) in diameter but are usually smaller; reduction in plant vigor; yellowing plants which wilt in hot weather

Cause: Nematode

Comments

Galls can appear as quickly as a month prior to planting; nematodes prefer sandy soils and damage in areas of field or garden with this type of soil is most likely

Management

Plant resistant varieties if nematodes are known to be present in the soil; check roots of plants mid-season or sooner if symptoms indicate nematodes; solarizing soil can reduce nematode populations in the soil and levels of inoculum of many other pathogens

Kale

Kale or borecole (*Brassica oleracea* Acephala Group) is a vegetable with green or purple leaves, in which the central leaves do not form a head. It is considered to be closer to wild cabbage than most domesticated forms. The species *Brassica oleracea* contains a wide variety of vegetables, including broccoli, cauliflower, collard greens, and brussels sprouts. The cultivar group Acephala also includes spring greens and collard greens, which are extremely similar genetically. *Pieris rapae* is one of the most well known pests of the plant.

Figure: *Curly kale*

Kale Plant Protection

Another fall project to foil the diseases of kale is working compost into the soil. Most people know that compost is a great natural fertilizer, but did you know that it also helps the soil drain freely?

Soil that can't drain freely stays wet too long, and many fungi thrive in wet soil. Working in compost in fall allows it plenty of time to combine with the soil so that it is ready to manage water more efficiently in spring.

Kale pests also overwinter in garden debris and soil. Expose the insects to the harsh conditions at the surface soil by turning the soil several times over fall and winter.

Eliminating Kale Pests

Identifying and eliminating some of the most common kale pests can go a long way in your kale plant protection program. Common garden pests affecting kale include:

- Aphids – Allow natural predator insects to do as much of the work of controlling these insects as possible. If you must use an insecticide, use a soap-based product or neem oil. You may have to spray several times.
- Flea beetles – Good fall cleanup and weed removal are your best bet in controlling these insects, which chew tiny holes in the leaves. If these kale pests find their way to your plants anyway, choose an insecticide that is labelled for use against flea beetles and make sure the label states that it safe to use on kale.
- Caterpillars – You'll probably notice moths darting around the plant before you see the caterpillars. In most cases, you can hand pick them. In severe cases, or if you can't bear to touch them, you can use *Bacillus thuringiensis*, or Bt.

- Whiteflies – These tiny, whiteflying insects rise in a cloud above the plant. Use insecticidal soap or neem oil and spray every few days until they are gone.

Kale Disease Prevention

Starting a kale disease prevention program will help eliminate most diseases of kale in the garden. Begin protecting kale plants by implementing these control measures:

- Water the soil rather than the plant. Wet plants are more susceptible to diseases than dry ones.

Also, avoid splashing soil onto the plants when you water.

- Clean your tools thoroughly before moving from one part of the garden to another. Don't forget your shoes! Bits of soil that travel from one part of the garden to another on the soles of your shoes may carry disease organisms.
- If you think your kale is infected with a disease, cut back on high-nitrogen fertilizers until you have it under control.
- Fungicides containing copper may help prevent disease infection or slow its progress, but they don't cure diseases. By using fungicides early, you may be able to hold off the disease until after you harvest your crop.

Mustard

Mustard plant are any of several plant species in the genera *Brassica* and *Sinapis*. Mustard seed is used as a spice. Grinding and mixing the seeds with water, vinegar or other liquids, creates the yellow condiment known as mustard.

The seeds can also be pressed to make mustard oil, and the edible leaves can be eaten as mustard greens.

Although some varieties of mustard plants were well-established crops in Hellenistic and Roman times, Zohary and Hopf note: "There are almost no archeological records available for any of these crops." Wild forms of mustard and its relatives the radish and turnip can be found over west Asia and Europe, suggesting their domestication took place somewhere in that area.

However, Zohary and Hopf conclude: "Suggestions as to the origins of these plants are necessarily based on linguistic considerations."

Varieties

*Mild white mustard (*Sinapis hirta*) grows wild in North Africa, the Middle East and Mediterranean Europe, and has spread farther by long cultivation; oriental mustard (*Brassica juncea*), originally from the foothills of the Himalaya, is grown commercially in India, Canada, the UK, Denmark and the US; black mustard (*Brassica nigra*) is grown in Argentina, Chile, the US and some European countries. Canada and Nepal are the world's major producers of mustard seed, between them accounting for around 57% of world production in 2010.*

There has been recent research into varieties of mustards that have a high oil content for use in the production of biodiesel, a renewable liquid fuel similar to diesel fuel. The biodiesel made from mustard oil has good cold flow properties and cetane ratings. The leftover meal after pressing out the oil has also been found to be an effective pesticide.

An interesting genetic relationship between many species of mustard has been observed, and is described as the Triangle of U.

Integrated Pest Management (IPM): Rapeseed-mustard crops are highly vulnerable to insect-pests at different stages of growth. Of which mustard aphid (Lipaphis erysimi) is the key pest followed by white mustard saw fly (Athalia lugens proxima), painted bug (Bagrada crusiferarum), pea leaf miner and bihar hairy caterpillar (Diacrisia obliqua).

Diseases to a large extent are responsible for low and unstable production of rapeseed-mustard and cause the yield losses ranging from 10-90%. The major diseases are White rust, Alternaria blight, and downy mildew and minor diseases are powdery mildew, bacterial rot and Sclerotinia stem rot.

Integrated pest management package for rape seed-mustard crop of Sikkim, recommended by the Central Integrated Pest Management Centre, Gangtok, Sikkim is as follows:

Pest Monitoring: The objective of pest monitoring is to detect the initial development of pest and disease and also bio-control potentials in the field situation. For this the following operations should be done:

Field Surveys: Survey, surveillance and monitoring of all the important pest should be done regularly i.e. weekly in general and twice a week during flowering/pod formation stage.

Field Scounting: Based on the observations, the farmers at village level have to be mobi-with green colour, diseased plant/leaves with yellow colour. While drawing the pest and the defenders the character should be taken to draw them at appropriate part of plant, where they are seen at the time of observations. The common name of pests and defenders and their population count should also be given along with diagram. The weather factors should be reflected in the chart by drawing the diagram of sun just above the plant it the attribute is sunny. If cloudy, the clouds may be drawn in place of sun. In case of partially sunny, the diagram of sun may be half marked with clouds.

Group Discussion and Decision Making: The observations recorded in the previous and current charts should be discussed among the farmers by raising questions relating to change in pest and defender population in relation to crop stage, soil condition, weather factor such as rainy, cloudy or sunny etc. The group may evolve a strategy based on weekly AESA, ETL and corresponding change in Pest: Defender ratio and take judicious decision for specific pest management practices.

Economic Threshold Levels (ETLS): To determine ETL of mustard aphid, observe 30-40 widely scattered plants in a hectare or per location or per spot, twice a week and then count the number of aphids (nymphs and adults) or measure the colony size (in centimetre) on 10 cm terminal portion of the central shoot or observe 250 plants/ha selected randomly from all the sides of the field to work the per cent plants infested.

Pest & Economic Threshold Levels (ETLs): Mustard Aphid: 20-50 aphids/10 cm terminal portion of the central shoot or 0.5-1.0 cm colony size on terminal portion of the central shoot or 30-40% aphid colonies or 30-40% plants infested by the aphid colonies (mother aphid with few young ones).

Integrated Pest Management Strategies

1. Deep summer ploughing wherever possible;
2. Destruction of plant debris of previous crops;
3. Grow resistant/tolerant varieties;
4. Early sowing of crop to escape the mustard aphid and major disease damage;
5. Use recommended doses of fertilizers to ensure adequate vigour in plants;
6. Irrigate the crop in 4th week after sowing to reduce the painted bug incidence, if present.

Mechanical Practices

1. Collection and destruction of early instar larvae/carterpillar;
2. Collection and destruction of egg masses;
3. Removal of aphid infested twigs in early stage.

Biological Control Practices: Conserve the natural bio-control agents namely coccinelids, chrysopids and syrphids by selecting the safe insecticides (endosulfan) safer for predators and pollinators, proper timing of spray (morning and evening hours) and following the need based application of insecticide/fungicide depending upon the economic threshold levels or disease appearance. The following are the defender of mustard crop:

1. ***Chrysopa:*** Green lacewing adults are delicate, light green insect with net like wings. The eggs are stalked. Its grubs possess sickle shaped mouth parts, wander on plants in search of soft bodied insects and prey voraciously on them.
2. ***Lady Bird Beetle:*** The lady bird beetles are predatory. The eggs are yellowish, larvae blackish and fast moving, pupae are also blackish and are resting stage. The adults are small in size yellowish in colour and having black spots. The adults and grubs feed on aphids.
3. ***Syrphid Fly:*** The larva is whitish in colour having sharp mouth parts which feed on aphids. The adult does not feed. These are also called hover flies and look like a slender housefly.

Integrated Pest management of important insect pests and diseases are as follows:

INSECT-PESTS:

- Grow improve varieties;

- Crop should be sown early at optimum time recommended for particular area. The crop sown before 5th October in north India often escape aphid damage;
- Use balanced dose of fertilizers. Over dosing of fertilizer particularly nitrogen would attract more aphids.
- Plucking and destruction of infested twigs is very useful. It should be done 2-3 times at 10 days interval early in the crop season;
- Spray the crop before the aphid reaches the economic –threshold level (ETL);

Mustard saw fly:

- Follow the crop sanitation practices;
- Timely irrigation helps in killing larvae through drowning.

Painted bug:

- Follow clean cultivation by weeding, hoeing and burn the debris around the field;
- Apply first irrigation 3-4 weeks after sowing wherever possible;
- Thresh the crop as early as possible to avoid further losses and dispose off plant material immediately.

Pea leaf miner:

- Pluck all the infested leaves and burn them to kill the larvae and pupae resting inside.

Bihar hairy caterpillar:

- Mechanical control is a non chemical eco-friendly method which provides an effective check by removing infested leaves containing first/second instar gregarious phase larvae and dip them in kerosenized water or insecticidal solution;
- Removal of alternative host plants from the surroundings is beneficial.

Broccoli

Broccoli is an edible green plant in the cabbage family, whose large flowering head is used as a vegetable. The word broccoli comes from the Italian plural of *broccolo*, which means “the flowering top of a cabbage”, and is the diminutive form of *brocco*, meaning “small nail” or “sprout”. Broccoli is often boiled or steamed but may be eaten raw.

Broccoli is classified in the Italica cultivar group of the species *Brassica oleracea*. Broccoli has large flower heads, usually green in colour, arranged in a tree-like structure on branches sprouting from a thick, edible stalk. The mass of flower heads is surrounded by leaves. Broccoli resembles cauliflower, which is a different cultivar group of the same species.

Broccoli is a result of careful breeding of cultivated leafy cole crops in the Northern Mediterranean in about the 6th century BC. Since the Roman Empire, broccoli has been considered a uniquely valuable food among Italians. Broccoli was brought to England from Antwerp in the mid-18th century by Peter Scheemakers. Broccoli was first introduced to the United States by Italian immigrants but did not become widely known there until the 1920s.

Varieties

Figure: *Broccoli plants in a nursery*

There are three commonly grown types of broccoli. The most familiar is Calabrese broccoli, often referred to simply as "broccoli", named after

Calabria in Italy. It has large (10 to 20 cm) green heads and thick stalks. It is a cool season annual crop. Sprouting broccoli has a larger number of heads with many thin stalks. Purple cauliflower is a type of broccoli sold in southern Italy, Spain, and the United Kingdom. It has a head shaped like cauliflower, but consisting of tiny flower buds. It sometimes, but not always, has a purple cast to the tips of the flower buds.

Other cultivar groups of *Brassica oleracea* include cabbage (Capitata Group), cauliflower and Romanesco broccoli (Botrytis Group), kale and collard greens (Acephala Group), kohlrabi (Gongylodes Group), Brussels sprouts (Gemmifera Group), and Chinese broccoli (Alboglabra Group). Rapini, sometimes called "broccoli raab" among other names, forms similar but smaller heads, and is actually a type of turnip (*Brassica rapa*). Broccolini or "Tenderstem broccoli" is a cross between broccoli and Chinese broccoli. Beneforté is a variety of broccoli containing 2-3 times more glucoraphanin that was produced by crossing broccoli with a wild Brassica variety, *Brassica oleracea* var *villosa.*

Cultivation

Broccoli is a cool-weather crop that does poorly in hot summer weather. Broccoli grows best when exposed to an average daily temperature between 18 and 23 °C (64 and 73 °F). When the cluster of flowers, also referred to as a "head" of broccoli, appear in the centre of the plant, the cluster is green. Garden pruners or shears are used to cut the head about an inch from the tip. Broccoli should be harvested before the flowers on the head bloom bright yellow.

While the heading broccoli variety performs poorly in hot weather, mainly due to insect infestation, the sprouting variety is more resistant, though attention must be paid to sucking insects (such as aphids), caterpillars and whiteflies. Spraying of *bacillus thuringiensis* can control caterpillar attacks, while a citronella vase may ward off whiteflies.

Pests

Mostly introduced by accident, "cabbage worms", the larvae of *Pieris rapae*, the small white butterfly are a common pest in broccoli.

As the most popular member of the cabbage family, broccoli is always in high demand at the table. Because it's a cool weather crop, you'll see your best results growing broccoli in the spring or fall. Florets that mature in autumn when nights turn chilly taste the sweetest. Broccoli seeds sprout best when soil temperatures range between 60 and 70 degrees Fahrenheit.

Broccoli Types to Try

Large-headed varieties produce the familiar domed heads that are composed of numerous clustered florets. Many large-headed varieties produce smaller side shoots after the primary head is harvested.

Sprouting varieties grow into bushier plants that produce numerous small heads. These varieties are at their best when grown from fall to spring in mild winter climates.

Romanesco varieties produce elegantly swirled heads composed of symmetrically pointed spirals. These large plants need plenty of space, excellent soil and good growing conditions to do well.

Broccoli raab is grown for its immature flower buds, which have a stronger flavour than regular broccoli. Broccoli raab (closely related to turnips) is popular in Asian and Italian cooking.

When to Plant Broccoli

For a summer harvest, start seeds indoors six weeks before your last spring frost, and set out hardened-off seedlings when they're about four weeks old. You can also seed broccoli directly into a nursery bed and transplant the seedlings to your garden. Direct-sow broccoli raab starting three weeks before your last frost.

For a fall harvest, start seeds indoors 12 to 14 weeks before your first fall frost, and set the seedlings out when they're four to six weeks old. Direct-sow broccoli raab starting eight weeks before your first fall frost.

How to Plant Broccoli

Broccoli is a heavy feeder, and plants take up nutrients best when the soil pH is between 6.0 and 7.0. Choose a sunny site with fertile, well-drained soil. Loosen the planting bed and mix in up to 1 inch of mature compost. Unless your soil is very fertile, also mix in a high-nitrogen organic fertilizer such as alfalfa meal or composted poultry manure. Water the bed thoroughly before setting out seedlings. Allow 18 to 20 inches between plants. Dwarf varieties can be planted 12 inches apart.

Broccoli Disease and Pest Prevention Tips

- Leaf-eating caterpillars — including army worms, cabbageworms, and cabbage loopers — like to feast on broccoli leaves. In summer, harlequin bugs and grasshoppers can devastate young plants. Prevent these problems by growing plants beneath row covers.
- When insect pressure is light, keep plants healthy by watching them closely and picking pests by hand. Weekly sprays with Bt

(*Bacillus thuringiensis*) or spinosad will control cabbageworms, the most serious broccoli pest.

- Plants that suddenly collapse may have been hit by cabbage root maggots, which are rice-sized fly larvae that feed on broccoli roots. In areas where this pest has been seen before, plant seedlings deeply, pressing the soil firmly around the stems. Prevent adults from laying eggs by covering the ground around each plant with a square of window screen or lightweight cloth.

Harvesting and Storing Broccoli

Harvest broccoli heads when the florets around the edges of the head begin to show slight loosening, but when the beads in most of the crown are still tight.

Cut the stems at an angle — this will keep water from pooling inside the cut stem and causing rot. Refrigerate the cut heads immediately. To get top home-preserved quality, steam-blanch broccoli before freezing it. Continue to water the plants after harvest, as most varieties will produce secondary heads.

Watch broccoli raab closely, and harvest just as the first flowers show their yellow petals. Sprouting broccoli and broccoli raab are also cut-and-come-again crops that produce a second flush of buds after the first ones have been harvested.

When plants are spaced 18 inches apart, average yields are about 1 pound of broccoli per foot of row. Three to four plants per person is sufficient for fresh summer eating, but you should triple that number if you want a freezer crop for winter.

Saving Broccoli Seeds

Most broccoli varieties must be exposed to winter chilling before they will flower heavily, but plants die if exposed to single-digit temperatures. Because of this, most seed is produced in mild winter areas.

Isolate plants to keep broccoli from crossing with cabbage, kale, and other close cabbage cousins. Broccoli raab will cross with turnips.

Harvest seeds when the slender pods dry to tan and the seeds inside are dark brown or black. Select the largest, most perfect seeds for planting.

Broccoli seeds will keep for five years when stored in cool, dry, dark conditions. Test seeds that are more than three years old before relying on them for a main crop — just put a few in a wet paper towel for five days to confirm they will sprout.

Watercress

Watercress, with the botanical name *Nasturtium officinale*, is a fast-growing, aquatic or semi-aquatic, perennial plant native to Europe and Asia, and one of the oldest known leaf vegetables consumed by humans. It is a member of the family Brassicaceae, botanically related to garden cress, mustard and radish — all noteworthy for a peppery, tangy flavour.

The hollow stems of watercress are floating, and the leaves are pinnately compound. Small, white and green flowers are produced in clusters.

Figure: *Watercress (*Nasturtium officinale)

Taxonomy

Nasturtium nasturtium-aquaticum *(nomenclaturally invalid) and* Sisymbrium nasturtium-aquaticum *L. are synonyms of* N. officinale. *Watercress is also listed in some sources as belonging to the genus* Rorippa, *although molecular evidence shows the aquatic species with hollow stems are more closely related to* Cardamine *than* Rorippa. *Despite the Latin name, watercress is not particularly closely related to the flowers popularly known as nasturtiums (*Tropaeolum majus*);* T. majus *belongs to the family Tropaeolaceae, a sister taxon to the Brassicaceae within the order Brassicales.*

Cultivation

Cultivation of watercress is practical on both a large-scale and a garden-scale. Being semi-aquatic, watercress is well-suited to hydroponic cultivation, thriving best in water that is slightly alkaline. It is frequently produced around the headwaters of chalk streams. In many local markets, the demand for hydroponically grown watercress exceeds supply, partly because cress leaves are unsuitable for distribution in dried form, and can only be stored fresh for a short period.

Watercress can be sold in supermarkets in sealed plastic bags, containing a little moisture and lightly pressurised to prevent crushing of contents. This has allowed national availability with a once-purchased storage life of one to two days in chilled/refrigerated storage.

Also sold as sprouts, the edible shoots are harvested days after germination. If unharvested, watercress can grow to a height of 50–120 centimetres (1.6–3.9 ft). Like many plants in this family, the foliage of watercress becomes bitter when the plants begin producing flowers.

Distribution

In some regions, watercress is regarded as a weed, in other regions as an aquatic vegetable or herb. Watercress has been grown in many locations around the world.

In the United Kingdom, watercress was first commercially cultivated in 1808 by the horticulturist William Bradbery, along the River Ebbsfleet in Kent. Watercress is now grown in a number of counties of the United Kingdom, most notably Hertfordshire, Hampshire, Wiltshire and Dorset. The town of Alresford, near Winchester, holds a Watercress Festival that brings in more than 15,000 visitors every year, and a preserved steam railway line has been named after the local crop. In recent years, watercress has become more widely available in the UK, at least in the southeast; it is stocked pre-packed in some supermarkets, as well as fresh by the bunch at farmers' markets and greengrocers.

In the United States in the 1940s, Huntsville, Alabama was locally known as the "watercress capital of the world". Today, Oviedo, Florida in the United States is known by that title, while Alresford in England is considered to be that nation's watercress capital

Pests and Diseases of Watercress

With regards to watercress, it's worth noting that they don't suffer too much from pests or disease but should you be unlucky enough for your plants to become infected by a fungal infection, it is likely only to occur at the propagation stage rather than on mature plants in their aquatic environment - with the exception of crook root. Therefore, it is particularly important to apply good high standards of hygiene during the propagation stage in order to minimise the need for subsequent chemical use. Be aware that watercress diseases tend to be most prevalent during dull, cool weather when the crop is unable to grow away.

Damping Off (Pythium Species)

Cause: This is very common fungal infection which is implicated in the disease commonly known as 'damping off'. When propagating seed in damp soil, the soft underground plant tissues can be attacked particularly if conditions are extremely and if plant growth is slow. Any aerial parts of the plant should remain unaffected unless they have contact with the soil.

Symptoms: You will typically lose part – or in extreme cases – all of the root system and your plants (rather obviously) will begin to lack vigour. Some Pythium species only attack the root tips, which can turn from white to brown and ending with their eventual collapse.

Treatment: Washing and disinfecting seed trays will help to minimise disease risk. Compost should always be kept from sources that may have been exposed to the fungus, e.g. by contaminated dust, or water , and your watering supply must also be kept free from sources of infection.

Crook Rot

Cause: This destructive fungal disease invades the plants by means of spores that penetrate the root cells. This fungus then goes on to systematically invade the leaves and stems of young shoots. The organism will also multiply within the root, producing large numbers of viable spores. At certain stages during its life cycle, only dormant spores are produced which are highly resistant to poor environmental conditions. When favourable conditions return, the dormant spores become active, and become released into the environment where it can infect more host plants. Crook root is especially most damaging in winter when watercress is growing more slowly.

Symptoms: Once the fungus reaches the new growth, the plant will begin to produce swollen and malformed, leaves and stems. As the disease moves into the secondary shoots, these will become stunted resulting in malformed plants.

Control: There are no chemicals that can be used to control this pathogen, however you can try purging your watercress with large volumes of water as this can reducing the rate of infection by washing away the viable spores.

Flea Beetles

Cause: This is perhaps the most recognised pest of watercress due to the immense, characteristic damage that these beetles can cause.

Symptoms: These beetles can cause significant damage by leaving copious amounts of small holes in the leaves. This normally serious infestation is usually experienced at two distinct times of the year, usually in April and July.

Control: Beetles can be removed by submerging your crop of watercress for about 2 hours. This causes the beetles to float off, at which point they can be skimmed off the surface of the water. You can also consider "trap crops" such as radish which may help lure the flea beetles away from your watercress. With the appropriate equipment – and if you don't have too many plants - Flea beetles can be vacuumed off the foliage. However this must be repeated frequently to avoid another invasion of your plants.

Mustard Beetle (Phaedon Cochleariae)

Causes: Like flea beetles, adult mustard beetles can lead to an almost continuous infestation of watercress throughout the summer.

Symptoms: Eggs are not laid on Watercress so damage is limited to the leaf and by the feeding adult beetles.

Control: Again, these beetles can be removed by flooding the cropping beds for about 2 hours to allow them to naturally float off before being skimmed off the surface of the water.

Aphids

Cause: Most aphid infestations are caused by the black bean aphid or the Peach blossom aphid (Myzus persicae).

Symptoms: Aphids generally feed on watercress over the summer, but severe infestations may cause serious losses. They damage and weaken the plant by sucking the sap out of pressurised parenchyma cells found just below the leaf's surface. Clusters of these small insects are readily identifiable, normally at the plants tips or on the underside of their leaves. In severe cases, the infected parts can begin to curl down due to the quantity of sap being removed from that area. Aphids will also excrete honeydew which causes leaves to first become sticky and shiny. They eventually turn black because of a sooty-mould fungus growth.

Control: There is no chemical control that is recommended against use on watercress, however using trap crops such as lettuce or bedding nasturtiums may help to entice aphids to these alternatives rather than to your edible crop.

3

Root Crops and Cucurbits

Root crops include those well known favourites such as carrots, parsnips, beetroot, radish, turnip, kohlrabi and some others you may not have tried growing such as salsify and celeriac.

Carrots

The carrot is a variable biennial plant, usually growing up to 1 m tall and flowering from June to August.

The umbels are claret-coloured or pale pink before they open, then bright white and rounded when in full flower, measuring 3–7 cm wide with a festoon of bracts beneath; finally, as they turn to seed, they contract and become concave like a bird's nest.

The dried umbels detach from the plant, becoming tumbleweeds.

Similar in appearance to the deadly poison hemlock, *D. carota* is distinguished by a mix of bi-pinnate and tri-pinnate leaves, fine hairs on its stems and leaves, a root that smells like carrots, and occasionally a single dark red flower in its centre.

Modern carrots are believed to have originated in Afghanistan and were introduced into Europe in 800-1000 AD.

The carrot is essentially a biennial, grown as an annual. It is the most popular root vegetable, because it is productive, nutritious and easily grown.

Carrots are an excellent source of beta- carotene, an important antioxidant from which the body makes Vitamin A.

Varieties

Carrots have come in many shapes and sizes, from long and tapering, to short and stumpy. Colours can range from white to cream and yellow, to orange, red and even purple. Most modern varieties are various shades of orange.

Long: 'Western Red' has long tapering roots, 30 cm or more long with strong tops. 'Majestic' is a slightly shorter improved form, with less brittle roots and which resists bolting to seed.

Medium: 'Topweight' grows up to 25 cm long, is slightly tapered with a stumpy tip. 'All Seasons' is similar, but up to 30 cm long, more cylindrical and excellent for year round sowing. 'Tip Top' is 25 cm long, smooth and cylindrical with excellent flavour.

Short: Chantenay types like 'Early Chantenay' and 'Chantenay Long' are short, 17-20 cm long and tapered with a stumpy tip. They are good for planting in shallower, heavier soils.

Baby: 'Amsterdam Forcing', 'Baby' and others produce small roots 10-15 cm long with fine texture and flavour.

Round: 'Paris Market' and 'French Round' produce almost spherical roots 3-5 cm across.

Growing Requirements

Carrots are adaptable and will grow under a range of climates, with best results in cool to mild conditions. Deep loamy soil or sandy loam is preferred, but well-prepared clay loam also grows quality carrots. If the bed has been well prepared, with organic matter and fertiliser for a previous crop, carrots can be grown without further preparation. If not, dig the soil well, to 30 cm or so deep. Organic matter should be old and fine. Avoid adding large quantities of coarse decomposing organic materials immediately before sowing – this is a major cause of hairy and forking roots.

Sowing and Growing

Establish carrots in frost free areas between August and March, and in colder areas between September and February. Two sowings of about one square metre in spring and again in late summer or early autumn will provide a steady supply of carrots. Carrot seed is not long-lived, so use fresh seed each year. Always check packet instructions for correct sowing times.

Sow seed in drills or rows and thin out seedlings later. The spacing after thinning depends on the desired result. Where baby or small carrots are desired, sow fairly thickly and leave unthinned or thin to 1 cm or so. If you want large roots, thin to 5-10 cm apart. Another method is to thin initially to 1 cm or so. When roots reach 1-2 cm at their tops, harvest alternate plants and use them as baby carrots. Then, leave the remainder to grow to full size. Mark out rows about 25-30 cm apart.

If fertiliser is needed, scatter it along the row. Form a shallow drill or furrow up to 1 cm deep with a rake handle or garden stake. Sow seed thinly and cover to less than 1 cm deep. Mulch lightly with dry grass clippings or similar to conserve moisture. During hot summer weather, use damp bags over the bed for the first week to help reduce the soil temperature. Water lightly to keep moist until seedlings emerge. Shadecloth suspended over young seedlings will protect them from the hot sun for the first few weeks after they emerge.

Watering

Once seedlings are established, watering should be deep and relatively infrequent to encourage roots to grow straight and deep. Two or three times a week in the early stages of growth. Once or twice a week in summer is sufficient in an established crop. Mulch between the rows.

Harvest Roots as Required

Carrots keep well in the ground. Late-sown carrots tend to become woody and bolt to seed if left in until the spring. Roots will store well in the refrigerator in vegetable bags.

Problems

Carrots are relatively less affected by pests and diseases than many other vegetables. A sunny site with well aerated soil will avoid most problems.

Nematodes

These microscopic worm- like animals in the soil cause wart-like lumps on the roots and unthrifty plants.

Vegetable Weevil

Both adults and larvae are about 12 mm long, and can damage tops and roots by extensive tunnelling. They are active mostly during winter, so young autumn sown crops are the most susceptible to infestation. They also favour other root vegetables, brassicas, lettuce and silverbeet. Control by rotating with non host crops like peas, beans or sweet corn and control weeds, particularly capeweed, marshmallow and cudweed.

Diseases

Under wet conditions some rotting and leaf spotting diseases may occur but this is not normally a problem.

Aphids

The most common is the carrot aphid, most prevalent in the milder weather of spring and autumn. Infestations can cause curling and twisting of leaves. The carrot aphid transmits the incurable virus disease Carrot Motley Dwarf from which plants become yellow, stunted and unproductive. Control aphids with sprays of pyrethrum or malathion.

Parsnips

The parsnip (*Pastinaca sativa*) is a root vegetable closely related to the carrot and parsley. It is a biennial plant usually grown as an annual. Its long tuberous root has cream-coloured skin and flesh and can be left in the ground when mature as it becomes sweeter in flavour after winter frosts. In its first growing season, the plant has a rosette of pinnate, mid-green leaves. If unharvested, it produces its flowering stem, topped by an umbel of small yellow flowers, in its second growing season. By this time the stem is woody and the tuber inedible. The seeds are pale brown, flat and winged.

The parsnip is native to Eurasia. It has been used as a vegetable since antiquity and was cultivated by the Romans, although there is some confusion in the literature of the time between parsnips and carrots. It was used as a sweetener before the arrival in Europe of cane sugar. It was introduced into the United States in the nineteenth century.

The parsnip is usually cooked but can also be eaten raw. It is high in vitamins and minerals, especially potassium. It also contains antioxidants and both soluble and insoluble dietary fibre. It can be cultivated in deep, stone-free soils and is attacked by the carrot fly and other insect pests, viruses and fungal diseases, of which canker is the most serious. In sunlight, handling the stems and foliage can cause a skin rash.

Description

Parsnips are grown for their fleshy, edible cream-coloured taproots. The roots are generally smooth, although lateral roots sometimes form. Most are cylindrical, but some cultivars have a more bulbous shape, which generally tend to be favoured by food processors as they are more resistant to breakage. The plant has a apical meristem that produces a rosette of pinnate leaves, each with several pairs of leaflets with toothed margins. The lower leaves have short stems, the upper ones are stemless, and the terminal leaves have three lobes. The highly branched floral stem is hollow and grooved, and can grow to more than 150 cm (60 in) tall.

Parsnip is a biennial with a rosette of roughly hairy leaves that has a pungent odor when crushed. The petioles are grooved and have sheathed bases. The leaves are once- or twice-pinnate with broad, ovate, sometimes lobed leaflets with toothed margins; they grow up to 40 cm (16 in) long.

The flower stalk develops in the second year, growing to a height of 40 to 200 cm (20 to 80 in). It is hairy, grooved, hollow (except at the nodes), and sparsely branched. It has a few stalkless, single-lobed leaves measuring 5 to 10 cm (2 to 4 in) long that are arranged in opposite pairs.

The yellow flowers are in a loose, compound umbel measuring 10 to 20 cm (4 to 8 in) in diameter. There are 6–25 straight pedicels, each measuring 2–5 cm (1–2 in) that support the umbellets (secondary umbels). The umbels and umbellets usually have no upper or lower bracts. The flowers have tiny sepals or lack them entirely, and measure about 3.5 mm. They consist of five yellow petals that are curled inward, five stamens, and one pistil. The fruits, or schizocarps, are oval and flat, with narrow wings and short, spreading styles. They are coloured straw to light brown, and measure 4–8 mm long.

*Despite the slight morphological differences between the two, wild parsnip is the same taxon as the cultivated version, and the two will readily cross-pollinate. Parsnip has a chromosome number of 2*n*=22.*

History

Like carrots, parsnips are native to Eurasia and have been eaten there since ancient times. Zohary and Hopf note that the archaeological evidence for the cultivation of the parsnip is "still rather limited", and that Greek and Roman literary sources are a major source about its early use. They warn that "there are some difficulties in distinguishing between parsnip and carrot (which, in Roman times, were white or purple) in classical writings since both vegetables seem to have been sometimes called pastinaca yet each vegetable appears to be well under cultivation in Roman times".

The parsnip was much esteemed and the Emperor Tiberius accepted part of the tribute payable to Rome by Germany in the form of parsnips.

In Europe, the vegetable was used as a source of sugar before sugar cane and beet were available. As pastinache comuni, the "common" pastinaca figures in the long list of comestibles enjoyed by the Milanese given by Bonvesin de la Riva in his "Marvels of Milan" (1288). This plant was introduced to North America simultaneously by the French colonists in Canada and the British in the Thirteen Colonies for use as a root vegetable, but in the mid-nineteenth century it was replaced as the main source of starch by the potato and consequently was less widely cultivated.

In 1859, a new cultivar called "Student" was developed by James Buckman at the Royal Agricultural College in England. He back-crossed cultivated plants to wild stock, aiming to demonstrate how native plants could be improved by selective breeding. This experiment was so successful that the Student became the major variety in cultivation in the late nineteenth century.

Taxonomy

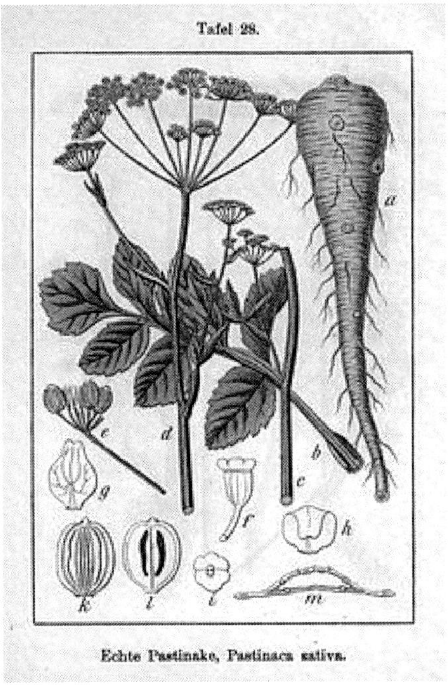

Figure: *Illustration from Johann Georg Sturm's 1796* Deutschlands Flora in Abbildungen

In Eurasia, some authorities distinguish between cultivated and wild versions of parsnip by using subspecies *sylvestris* for the latter, or even elevating it to species status as *Pastinaca sylvestris*. In Europe, various subspecies have been named based on characteristics such as the hairiness of the leaves, the extent to which the stems are angled or rounded, and the size and shape of the terminal umbel.

The etymology of the genus name *Pastinaca* is not known with certainty, but is probably derived from either the Latin word *pastino*, meaning "to prepare the ground for planting of the vine" or *pastus*, meaning "food". The specific epithet *sativa* means "sown".

Uses

Parsnips resemble carrots and can be used in similar ways but they have a sweeter taste, especially when cooked. While parsnips can be eaten raw, they are more commonly served cooked. They can be baked, boiled, pureed, roasted, fried or steamed. When used in stews, soups and casseroles they give a rich flavour.

In some cases, the parsnip is boiled and the solid portions are removed from the soup or stew, leaving behind a more subtle flavour than the whole root, and starch to thicken the dish. Roast parsnip is considered an essential part of Christmas dinner in some parts of the English-speaking world and frequently features in the traditional Sunday Roast.

Parsnips can also be fried or thinly sliced and made into crisps. Parsnips can be made into a wine that has a taste similar to Madeira.

In Roman times, parsnips were believed to be an aphrodisiac. However, parsnips do not typically feature in modern Italian cooking. Instead, they are fed to pigs, particularly those bred to make Parma ham.

Cultivation

The wild parsnip from which the modern cultivated varieties were derived is a plant of dry rough grassland and waste places, particularly on chalk and limestone. Parsnips are biennials but are normally grown as annuals. Sandy and loamy soils are preferable to silt, clay and stony ground as the latter produce short, forked roots.. Parsnip seed significantly deteriorates in viability if stored for long. Seeds are usually planted in early spring, as soon as the ground can be worked to a fine tilth, in the position where the plants are to grow.

The growing plants are thinned and kept weed free. Harvesting begins in late fall after the first frost, and continues through winter. The rows can be covered with straw to enable the crop to be lifted during frosty weather.

Low soil temperatures cause some of the starches stored in the roots to be converted into sugars, giving them a sweeter taste.

Cultivation Problems

Parsnip leaves are sometimes tunnelled by the larvae of the celery leaf miner (*Euleia heraclei*). Irregular pale brown passages can be seen meandering between the upper and lower surfaces of the leaves. The effects are most serious on young plants as whole leaves may shrivel and die. Treatment is by removing affected leaflets or whole leaves, or by chemical means.

The crop can be attacked by larvae of the carrot fly (*Chamaepsila rosae*). This pest feeds on the outer layers of the root, burrowing its way inside later in the season. Seedlings may be killed while larger roots are rendered useless. The damage done provides a point of entry for fungal rots and canker.

The fly is attracted by the smell of bruised tissue and damage can be minimized by sowing thinly, providing good growing conditions, avoiding thinning the plants and weeding carefully. Parsnip is used as a food plant by the larvae of some Lepidopteran species including the parsnip swallowtail, the common swift moth, the garden dart moth and the ghost moth. The larvae of the parsnip moth (*Depressaria radiella*), native to Europe and accidentally introduced to North America in the mid-1800s, construct their webs on the umbels, feeding on flowers and developing seeds.

Parsnip canker is a serious disease of this crop. Black or orange-brown patches occur around the crown and shoulders of the root accompanied by cracking and hardening of the flesh. It is more likely to occur when seed is sown into cold, wet soil, the pH of the soil is too low or the roots have already been damaged by carrot fly larvae. Better results may be obtained by choosing a resistant variety and sowing later in the season, producing smaller, more closely spaced plants. Several fungi are associated with canker, including *Phoma complanata*, *Ilyonectria radicicola*, *Itersonilia pastinaceae*, and *I. perplexans*.

In Europe, *Centrospora acerina* has been found to cause a black rot that kills the plant early. Watery soft rot, caused by *Sclerotinia minor* and *S. sclerotiorum*, causes the taproot to become soft and watery. A white or buff-coloured mold grows on the surface. The pathogen is most common in temperate and subtropical regions that have a cool wet season.

Violet root rot caused by the fungus *Helicobasidium purpureum* sometimes affects the roots, covering them with a purplish mat to which soil particles adhere. The leaves become distorted and discoloured and the mycelium can spread through the soil between plants. Some weeds can harbour this fungus and it is more prevalent in wet, acid conditions. Affected plants should be removed and burnt and susceptible crops should not be grown in the area for four years.

Erysiphe heraclei causes a powdery mildew that can cause significant crop loss. Infestation by this causes results in yellowing of the leaf and loss of foliage. Moderate temperatures and high humidity favour the development of the disease.

Several viruses are known to infect the plant, including seed-borne strawberry latent ringspot virus, parsnip yellow fleck virus, parsnip leafcurl virus, parsnip mosaic potyvirus, and potyvirus celery mosaic virus. The latter causes clearing or yellowing of those areas of the leaf immediately aligning the veins, the appearance of ochre mosaic spots, and leaf crinkling in infected plants.

Dangers

While the root of the parsnip is edible, handling the shoots and leaves of the plant requires caution as the sap is toxic. Like many other members of the family Apiaceae, the parsnip contains furanocoumarin, a photosensitive chemical that causes a condition known as phytophotodermatitis.

The condition is a type of chemical burn rather than an allergic reaction, and is similar to the rash caused by poison ivy. Symptoms include redness, burning, and blisters. Afflicted areas can remain discoloured for up to two years. Although there have been some reports of gardeners experiencing toxic symptoms after coming into contact with foliage, these have been small in number compared to the number of people that grow the crop.

The problem is most likely to occur on a sunny day when gathering foliage or pulling up old plants that have gone to seed. The symptoms have mostly been mild to moderate. The toxic properties of parsnip extracts are resistant to heating, or a storage period of several months. Toxic symptoms can also affect livestock and poultry in parts of their bodies where their skin is exposed.

Beetroot

*The beetroot, also known in North America as the table beet, garden beet, red or golden beet, or informally simply as the beet, is any of the cultivated varieties of beet (*Beta vulgaris*) grown for their edible taproots, especially* B. vulgaris *L. subsp.* conditiva. *They are among the most commonly encountered varieties in North America, Central America, and Europe.*

Varieties

Red globe-shaped varieties include 'Crimson Globe', 'Detroit Dark Red' and others. 'Baby' beets are 5 cm across. 'Cylindrica' has cylindrical roots up to 25 cm long. Others include 'Golden', 'White' and 'Italian Chioggia' with concentric rings of red and white.

Sowing and Growing

Deep light, friable soils produce the best roots. Beetroot seed is best sown from August to March or September to February in colder areas. Sowings later than this will bolt to seed in spring. Sow seed 12 mm deep in rows 20-30 cm apart and thin to 7-10 cm . General care is as for other root crops. Pests and diseases are not prevalent on well-grown beetroot.

Root vegetables, like beets, prevent you from monitoring the vegetables because they are below ground, but the leaves often reflect the overall health of the plant. Changes or damage to the beet leaves clue you in to potential problems with the vegetables. Identifying the cause of the leaf damage allows you to treat the condition to grow healthy beets.

Pests

Garden pests typically attack the beet leaves and cause identifiable damage. Cutworms eat the stems and leaves at the soil level. Wood ash applied near the plants and paper collars around the stems discourage cutworms. Aphids cause small, shiny specks, curled leaves, leaf deformities and a yellowish tint. Insecticidal soap helps control aphids. Flea beetles and black beetles leave small holes where they have eaten leaves, most often on seedlings.

Using diatomaceous earth near the seedlings helps. Cabbage loopers eat irregular-shaped holes in the beet leaves. Pick these caterpillars that are green with yellow stripes from the plants to reduce damage. Partially eaten leaves that are webbed together are often caused by webworms, which you should remove and destroy. If you notice slime on the eaten leaves, slugs or snails could be to blame. Place a shallow bowl of beer, with the lip at soil level, to drown slugs and snails. Leafminer larvae damage the leaves from the inside by tunneling through them. Get rid of leaves that show signs of tunneling or trails. Floating row covers also keep leafminers away.

Virus

Viruses that damage beets often cause visible changes to the leaves. Curly top virus causes leaves the curl and pucker while turning the veins purple. The beet plant growth often slows and the leaves become brittle or thick. Leafhoppers spread the virus, so keep these pests under control. Get rid of beets affected by curly top virus to prevent its spread.

Mosaic virus also stunts growth of the beets while turning the leaves brittle, distorted and covered with a mottled design in various green shades. Leafhoppers and aphids spread the disease through beet crops, so pest control is essential. Dig and dispose of diseased beets immediately to save the rest of the crop.

No Matches Found. Please try your search again.

Fungus

Beets are susceptible to certain fungal diseases that damage the leaves. Cercospora leaf spot occurs in warm, wet climates and causes purple margins or round spots with brown centres. Remove the damaged leaves and don't water the beets from above. Downy mildew is another fungus that affects beets.

It causes irregular spots on the leaves. On the bottom of the leaves you'll notice mold or powder. Some beet plants are bred to resist downy mildew. Better air circulation and changing where you plant beets each year reduces the risk.

The fungal diseases are more prevalent around weeds or plant debris, so keep the area around beets clear.

Nutrient Deficiency

Beet leaves can reveal deficiencies in the soil that affect growth. If the leaves begin to crinkle, the tips die or the leaf margins turn red, the plants may be low on boron. A lack of boron in the soil occurs if it is either too acidic or too alkaline. If a soil pH test reveals levels outside the 6.0 to 7.0 range, the growing conditions aren't ideal for the beets. Add borax to increase the boron of the soil at a rate of 2 ounces for every 30 square yards of soil.

Radish

The radish (Raphanus sativus) is an edible root vegetable of the Brassicaceae family that was domesticated in Europe in pre-Roman times. They are grown and consumed throughout the world. Radishes have numerous varieties, varying in size, colour and duration of required cultivation time.

There are some radishes that are grown for their seeds; oilseed radishes are grown, as the name implies, for oil production. Radish can sprout from seed to small plant in as little as 3 days.

Figure: *Radishes*

Description of the Radish Plant

Radishes are round to cylindrical with a colour ranging from white to red. A longer root form, ideal for cooking, grows up to 15 cm (6 in) long, while the smaller, rounder form is typically eaten raw in salads. The flesh initially tastes sweet, but becomes bitter if the vegetable is left in the ground for too long. Leaves are arranged in a rosette, with sizes ranging from 10–15 cm (4–6 in) in small cultivars, to up to 45 cm (18 in) in large cultivars. They have a lyrate shape, meaning they are divided pinnately with an enlarged terminal lobe and smaller lateral lobes. The white flowers are borne on a racemose inflorescence.

*The radish is a diploid species, and has 18 chromosomes (2*n*=18).*

Cultivation

Figure: *Growing radish plants*

Radishes grow best in full sun and light, sandy loams with pH 6.5–7.0. They are in season from April to June and from October to January in most parts of North America; in Europe and Japan they are available year-round due to the plurality of varieties grown.

Summer radishes mature rapidly, with many varieties germinating in 3–7 days, and reaching maturity in three to four weeks. Harvesting periods can be extended through repeated plantings, spaced a week or two apart.

As with other root crops, tilling the soil to loosen it up and remove rocks helps the roots grow. However, radishes are used in no-till farming to help reverse compaction.

Most soil types will work, though sandy loams are particularly good for winter and spring crops, while soils that form a hard crust can impair growth. The depth at which seeds are planted affects the size of the root, from 1 cm (0.4 in) deep recommended for small radishes to 4 cm (1.6 in) for large radishes.

Radishes are a common garden crop in the U.S., and the fast harvest cycle makes them a popular choice for children's gardens.

In temperate climates, it is customary to plant radishes every two weeks from early spring until a few weeks before the first frost, except during periods of hot weather. In warm-weather climates, they are normally planted in the fall.

After harvest, radishes can be stored without loss of quality for two or three days at room temperature, and about two months at 0 °C (32 °F) with a relative humidity of 90–95%.

Companion Plant

Radishes serve as companion plants for many other species, because of their ability to function as a trap crop against pests like flea beetles. These pests will attack the leaves, but the root remains healthy and can be harvested later.

Popular in salads, radishes add a spicy flavour, and some types can also be used for cooking. Radishes are productive and quick to mature, with varieties for year-round sowing.

Radishes need well prepared soil, like other roots, and must be grown quickly with generous watering. Slow grown radishes often become tough, hot and pithy. Apply about 40-60 grams of fertiliser per square metre if desired, but no more. Check seed packets for sowing details appropriate to the variety.

Varieties

Broadly speaking, radishes can be categorized into four main types (summer, fall, winter, and spring) and a variety of shapes lengths, colours, and sizes, such as red, pink, white, gray-black or yellow radishes, with round or elongated roots that can grow longer than a parsnip.

Spring or Summer Radishes

Figure: *European radishes (*Raphanus sativus*)*

Figure: *Germinating radishes, 10 days old*

Sometimes referred to as European radishes or spring radishes if they are planted in cooler weather, summer radishes are generally small and have a relatively short 3–4 week cultivation time.

- The April Cross is a giant white radish hybrid that bolts very slowly.
- Bunny Tail is an heirloom variety from Italy, where it is known as 'Rosso Tondo A Piccola Punta Bianca'. It is slightly oblong, mostly red, with a white tip.

- Cherry Belle is a bright red-skinned round variety with a white interior. It is familiar in North American supermarkets.
- Champion is round and red-skinned like the Cherry Belle, but with slightly larger roots, up to about 5 cm (2 in), and a milder flavour.
- Red King has a mild flavour, with good resistance to club root, a problem that can arise from poor drainage.
- Sicily Giant is a large heirloom variety from Sicily. It can reach up to two inches in diameter.
- Snow Belle is an all-white variety of radish, similar in shape to the Cherry Belle.
- White Icicle or just Icicle is a white carrot-shaped variety, around 10–12 cm (4–5 in) long, dating back to the 16th century. It slices easily, and has better than average resistance to pithiness.
- French Breakfast is an elongated red-skinned radish with a white splash at the root end. It is typically slightly milder than other summer varieties, but is among the quickest to turn pithy.
- Plum Purple a purple-fuchsia radish that tends to stay crisp longer than average.
- Gala and Roodbol are two varieties popular in the Netherlands in a breakfast dish, thinly sliced on buttered bread.
- Easter Egg is not an actual variety, but a mix of varieties with different skin colours, typically including white, pink, red, and purple radishes. Sold in markets or seed packets under the name, the seed mixes can extend harvesting duration from a single planting, as different varieties may mature at different times.

Winter Varieties

Figure: *Daikon*

Black Spanish or Black Spanish Round occur in both round and elongated forms, and are sometimes simply called the black radish or known by the French name *Gros Noir d'Hiver*. It dates in Europe to 1548, and was a common garden variety in England and France during the early 19th century. It has a rough black skin with hot-flavoured white flesh, is round or irregularly pear shaped, and grows to around 10 cm (4 in) in diameter.

Daikon refers to a wide variety of winter radishes from Asia. While the Japanese name daikon has been adopted in English, it is also sometimes called the Japanese radish, Chinese radish, Oriental radish or *mooli* (in India and South Asia). Daikon commonly have elongated white roots, although many varieties of daikon exist. One well known variety is April Cross, with smooth white roots. *The New York Times* describes Masato Red and Masato Green varieties as extremely long, well suited for fall planting and winter storage. The Sakurajima daikon is a hot-flavoured variety which is typically grown to around 10 kg (22 lb), but which can grow to 30 kg (66 lb) when left in the ground.

Seed Pod Varieties

The seeds of radishes grow in siliques (widely referred to as "pods"), following flowering that happens when left to grow past their normal harvesting period. The seeds are edible, and are sometimes used as a crunchy, spicy addition to salads. Some varieties are grown specifically for their seeds or seed pods, rather than their roots. The Rat-tailed radish, an old European variety thought to have come from East Asia centuries ago, has long, thin, curly pods which can exceed 20 cm (8 in) in length. In the 17th century, the pods were often pickled and served with meat. The München Bier variety supplies spicy seed pods that are sometimes served raw as an accompaniment to beer in Germany.

Figure: *Radish fruits, also called pods*

Figure: *Radish seeds*

Pests

Although often unintentionally introduced, the larvae of *Pieris rapae*, the small white butterfly is known to pest on the leaves of the radish.

Mustard Saw Fly

Mustard saw fly is most destructive is most destructive insect of radish in our region. If not controlled in time causes serious damages to the crop. It attacks the crop on leaves.

The grub feeds on the leaves and pods. In severe cases complete defoliation occurs. Spray with Malathion to control disease.

Aphids

Aphids are most serious pest particularly in rainy season. Under sever infestation plants are completely devitalized. Leaves and shoots curl up, become yellowish and finally die. Spraying with Malathion 0.05 per cent check the attack of aphids.

Diseases

Radish is a short duration crop. Therefore, prompt care has to be taken to control diseases and pest. Alternate blight white rust is important diseases.

Alternaria blight attack on leaves seen in the form of small, yellowish, slightly raised lesions. It is more in rainy season. Spray crop with copper- oxychloride to control disease.

Figure: *Kohlrabi stem with leaves*

Kohlrabi, Knol Kohl

Kohlrabi (German turnip or turnip cabbage) (*Brassica oleracea* Gongylodes group) is an annual vegetable, and is a low, stout cultivar of cabbage. Kohlrabi can be eaten raw as well as cooked.

Description

Kohlrabi has been created by artificial selection for lateral meristem growth (a swollen, nearly spherical shape); its origin in nature is the same as that of cabbage, broccoli, cauliflower, kale, collard greens, and Brussels sprouts: they are all bred from, and are the same species as the wild cabbage plant (Brassica oleracea).

The taste and texture of kohlrabi are similar to those of a broccoli stem or cabbage heart, but milder and sweeter, with a higher ratio of flesh to skin. The young stem in particular can be as crisp and juicy as an apple, although much less sweet.

Figure: *A basket of kohlrabi*

Except for the *Gigante* cultivar, spring-grown kohlrabi much over 5 cm in size tend to be woody, as do full-grown kohlrabi much over perhaps 10 cm in size; the Gigante cultivar can achieve great size while remaining of good eating quality. The plant matures in 55–60 days after sowing. Approximate weight is 150 g and has good standing ability for up to 30 days after maturity.

There are several varieties commonly available, including White Vienna, Purple Vienna, Grand Duke, Gigante (also known as "Superschmelz"), Purple Danube, and White Danube. Colouration of the purple types is superficial: the edible parts are all pale yellow. The leafy greens can also be eaten.

The best quality comes when grown during the milder months of autumn and spring. Soil and growing requirements are similar to other brassicas, turnips and swedes.

Varieties

'Early Purple', 'White Vienna' and 'Purple Globe' are common standard varieties.

Planting and Growing

Sow from midsummer to early autumn then late winter to spring. Water as needed during the season. Quick growth gives the best quality.

Maturity and Harvesting

Harvest when at full size.Problems: As for European brassicas.

Kohlrabi Pests and Problems

The most common kohlrabi pest (as with many other members of the cabbage family) is the cabbage white butterfly and its spawn, the cabbage worm. In addition, cutworms and powdery mildew can also be issues. To avoid problems with cabbage worms, grow your kohlrabi under floating row covers so the butterflies can't lay their eggs on your plants.

Turnip and Swede

The turnip or white turnip (*Brassica rapa* subsp. *rapa*) is a root vegetable commonly grown in temperate climates worldwide for its white, bulbous taproot. Small, tender varieties are grown for human consumption, while larger varieties are grown as feed for livestock.

In the north of England and Scotland, turnip (or *neep*; the word *turnip* itself is an old compound of *neep*) refers to the larger, yellow rutabaga root vegetable which is also known as the "swede" (from "Swedish turnip").

Figure: *Turnip roots*

Description

The most common type of turnip is mostly white-skinned apart from the upper 1–6 centimetres, which protrude above the ground and are purple, red, or greenish wherever sunlight has fallen. This above-ground part develops from stem tissue, but is fused with the root. The interior flesh is entirely white.

The entire root is roughly conical, but can be occasionally global, about 5–20 centimetres in diameter, and lacks side roots. The taproot (the normal root below the swollen storage root) is thin and 10 centimetres or more in length; it is trimmed off before marketing. The leaves grow directly from the above-ground shoulder of the root, with little or no visible crown or neck (as found in rutabagas).

Turnip leaves are sometimes eaten as "turnip greens" ("turnip tops" in the UK), and they resemble mustard greens in flavour. Turnip greens are a common side dish in southeastern US cooking, primarily during late fall and winter. Smaller leaves are preferred; however, any bitter taste of larger leaves can be reduced by pouring off the water from initial boiling and replacing it with fresh water.

Varieties specifically grown for the leaves resemble mustard greens more than those grown for the roots, with small or no storage roots. Varieties of B. rapa that have been developed only for the use of leaves are called Chinese cabbage. Both leaves and root have a pungent flavour similar to raw cabbage or radishes that becomes mild after cooking.

Turnip roots weigh up to about one kilogram, although they can be harvested when smaller. Size is partly a function of variety and partly a function of the length of time the turnip has grown. Most very small turnips (also called baby turnips) are speciality varieties. These are only available when freshly harvested and do not keep well. Most baby turnips can be eaten whole, including their leaves. Baby turnips come in yellow-, orange-, and red-fleshed varieties as well as white-fleshed. Their flavour is mild, so they can be eaten raw in salads like radishes and other vegetables.

As a root crop, turnips grow best in cool weather; hot temperatures cause the roots to become woody and bad-tasting. They are typically planted in the spring in cold-weather climates (such as the northern US and Canada) where the growing season is only 3–4 months. In temperate climates (ones with a growing season of 5–6 months), turnips may also be planted in late summer for a second fall crop. In warm-weather climates (7 or more month growing season), they are planted in the fall. 55–60 days is the average time from planting to harvest.

Turnips are a biennial plant, taking two years from germination to reproduction. The root spends the first year growing and storing nutrients, and the second year flowers, produces seeds, and dies. The flowers of the turnip are tall and yellow, with the seeds forming in pea-like pods. In areas with less than 7 month growing seasons, temperatures are too cold for the roots to survive the winter months. In order to produce seeds, it's necessary to pull the turnip and store it overwinter, taking care not to damage the leaves. During the spring, it may be set back in the ground to complete its life cycle.

Turnips and swedes are closely related root vegetables which are actually swollen stem bases. Both thrive in cool climates, so both are grown in Australia during the cooler months.

Varieties

'Purple Top White Globe' and 'White Stone'. 'Hakurai' is a Japanese turnip. Swede varieties include 'Champion Purple Top' and 'Royal Rose'.

Planting and Growing

Sow seed of turnip and swede from February to April in mild to warm areas. In cool to cold areas, seed can be sown from January to March, and a second crop from August to September. Sow seed 6-10 mm deep in rows 25-30 cm apart. Seedlings emerge in a week to 10 days. Thin to 8-10 cm spacings. General culture is similar to other root crops.

Harvesting and Storage

Turnips reach full size in 10-14 weeks and are best used fresh, soon after harvest. Swedes take longer to mature, at least 15 weeks and up to 20 weeks or more. Swedes maturing under cold, frosty conditions have the best flavour. Swedes keep much longer than turnips.

Pests and Their Control

Turnips are a staple of many food cultures and one of the most commonly grown root vegetables. Home gardeners love their flexible turnips for both their fleshy roots and leafy greens. Sadly, many bugs also enjoy a taste of homegrown turnips. Most turnip pests can be simply picked from the plants when present in small numbers, though a few may require more aggressive treatment.

Foliage-Feeding Caterpillars

Many foliage-feeding caterpillars have a taste for turnip greens. Beet armyworms, cabbage loopers, diamondback moth caterpillars and imported cabbageworms are common caterpillar pests. These caterpillars are often found feeding on leaf surfaces, eating from the outside edge inward. Identification isn't crucial, since all kinds of foliage-feeding caterpillars can be treated by handpicking or spraying the turnips with Bacillus thuringiensis. If many mixed types of caterpillars are present at once, you may choose to employ carbamate or a pyrethroid insecticide to hasten their elimination.

Cutworms

Cutworms are nocturnal, soil-based caterpillars that feed on the stems or low-hanging leaves of young plants. You may never see the pest, but you will know they've been active when seedlings are cut down at ground level, often in a row. Plow gardens thoroughly about 10 days before planting to destroy larvae and eggs, as well as any food sources. If cutworms are still especially bad, the insecticide indoxacarb can be employed.

No Matches Found. Please try your search again.

Aphids

Cabbage aphid (Brevicoryne brassicae), green peach aphid (Myzus persicae) and turnip aphid (Lipaphis erysimi) are active turnip feeders. These soft-bodied, tiny insects feed using sharp mouth parts that tap directly into the plant's tissues. Unless your homegrown turnips are being grown for market, aphids can often be tolerated. They may cause leaves to curl or appear wilted but rarely harm the turnip itself. When treatment is necessary, apply insecticidal soap when the temperature is under 90 degrees Fahrenheit and the weather is cloudy, since turnips can be sensitive to the sun shortly after treatment.

Whiteflies

Whiteflies are another sap-feeding insect that feeds upon turnips, though unlike aphids, they can and do fly readily when disturbed. These tiny insects resemble moths when in flight because of the reflective waxy coating on their wings. Whiteflies produce copious amounts of honeydew, which attracts sooty mold that can interfere with photosynthesis in turnips. Treat whiteflies when you see them with insecticidal soaps or horticultural oils. Repeat treatments as necessary, since whiteflies reproduce quickly.

Maggots

Cabbage and seedcorn maggots are difficult to tell apart at first glance. Both are small, white, legless larvae that live underground and burrow into roots. The seedcorn maggot, however, rarely feeds on the enlarging turnip root, instead preferring to nibble at young plants. Cabbage maggots, however, are a different story. They often penetrate turnips in multiple areas and tunnel throughout, leaving the turnips open to invasion from soil-borne fungi. A permit is required to apply diazinon to turnip crops, but it will help to keep these destructive pests at bay.

Celeriac (Turnip-Rooted Celery)

Celeriac, often called turnip-rooted celery, is grown mainly for its swollen leaf bases. It has a true celery flavour and can be used raw in salads or cooked in soups and stews. The leaf stalks can also be used as for celery. Culture is similar to that for celery, but celeriac is more adaptable and hardier. Seedlings are rarely seen in nurseries so plants need to be raised from seed. Celeriac is a root vegetable with a bulbous hypocotyl. In the Mediterranean Basin and in Northern Europe, celeriac grows wild and is widely cultivated. It is also cultivated in North Africa, Siberia, Southwest Asia, and North America. In North America, the Diamant cultivar predominates. Celeriac originated in the Mediterranean Basin.

Planting and Growing

Plant from November to April, 30cm apart in rows 40cm apart, with the bulbous swelling of the base at ground level. General care is similar to celery, although celeriac is more drought tolerant. Cut off any side shoots that appear at the base of the plant. Blanch the swollen base by covering with straw or dry mulch.

Harvesting and Storage

Celeriac can be left in the ground longer than celery until the swollen roots reach a good size. Trim off the foliage and excess roots, wash and store in the refrigerator as for celery. Well grown celeriac stored properly will keep for up to four months.

Pests and Diseases of Celery and Celeriac

Luckily if you choose to grow celeriac then you will be pleased to know that it grows strongly once established and is affected by very few pests and diseases. Indeed its strong growing characteristics make it a good choice for those new to growing vegetables as it is a great tasting vegetable that requires little care or maintenance. The main celeriac pests are the celery fly, and slugs and snails.

Celery fly – These small flies lay eggs on the leaves, which in turn become maggots, typically these will appear on the plants in April and May. The first sign that the celeriac is affected by them is the white, semi-transparent marks start to appear on the leaves of the plants, in time these marks on the leaves will go brown, the leaves will then start to become distorted and start to shrivel up.

The celery fly can be treated by using pesticide sprays but they only work if applied if the problem is spotted early on. The easiest way to avoid the celery fly is to plant celeriac in late May or early June, when the celery fly is less of a problem.

Slugs and snails will attack celeriac just as they love to attack just about any vegetables in the garden, they can be treated by the use of slug pellets, which will need putting down every time it has rained heavily.

Or you can place slug traps around the plants, typically these attract the slugs and snails in to them by using beer as the bait!

The slugs and snails then drown in the beer. Crushed eggshells can also be placed around the plants as neither pest like crawling over them, the slug pellets and traps however will kill them, but the eggshells will merely deter them.

Leaf spot is a fungus that can develop on the celeriac leaves and it mainly develops in wet and warm conditions. It takes the form of brown marks developing on the leaves, starting as small brown spots they will spread and join together eventually killing the leaf.

The leaf spot can be prevented by removing any infected leaves from the celeriac plants as soon as they develop.

Cucurbits

Figure: *Hodgsonia male plant*

The Cucurbitaceae are a plant family, sometimes called the gourd family, consisting of over a hundred genera, the most important of which are:

- *Cucurbita* – squash, pumpkin, zucchini, some gourds
- *Lagenaria* – mostly non-edible gourds
- *Citrullus* – watermelon (*Citrullus lanatus*) and others
- *Cucumis* – cucumber (*Cucumis sativus*), various melons
- *Luffa* – common name also luffa

The plants in this family are grown around the tropics and in temperate areas, where those with edible fruits were among the earliest cultivated plants both in the Old and New World. The *Cucurbitaceae* family ranks among the highest of plant families for number and percentage of species used as human food.

The Cucurbitaceae consist of approximately 125 genera and 960 species, mainly in regions tropical and subtropical. All species are sensitive to frost. Most of the plants in this family are annual vines but there are also woody lianas, thorny shrubs, and trees (*Dendrosicyos*). Many species have large, yellow or white flowers. The stems are hairy and pentangular. Tendrils are present at 90° to the leaf petioles at nodes. Leaves are exstipulate alternate simple palmately lobed or palmately compound. The flowers are unisexual, with male and female flowers on different plants (dioecious) or on the same plant (monoecious). The female flowers have inferior ovaries. The fruit is often a kind of modified berry called a pepo.

Classification

Figure: *Seen on this image is a selection of cucurbits of the South Korean Genebank in Suwon*

The about 125 existent genera in Cucurbitaceae include 960 species. The following is the classification as given by Charles Jeffrey as of 1990. However, a 2011 based on genetics does not support this taxonomy with 2 subfamilies and 8 tribes , but rather 15 tribes, 5 of them new, consisting of 95 genera, not 121.

Cucurbita

Cucurbita (Latin for gourd) is a genus in the gourd family Cucurbitaceae native to and originally cultivated in the Andes and Mesoamerica. The *Cucurbita* genus is an important source of human food and is used for other purposes such as beverages, medicine, oil, and detergent. Some *Cucurbita* species were brought to Europe after the discovery of America and are now used in many parts of the world. The plants, referred to as *squash*, *pumpkin* or *gourd* depending on species, variety and local parlance, are grown for their edible fruits and seeds. Real (bottle-)gourds, used as utensils or vessels, belong to the genus *Lagenaria* and are native to Africa. *Lagenaria* are in the same family and subfamily as *Cucurbita*, but different tribes.

There are five domesticated species: *Cucurbita argyrosperma*, *C. ficifolia*, *C. maxima*, *C. moschata*, and *C. pepo*. *C. pepo* includes varieties of both winter squash and summer squash, and *C. moschata* can be used as winter squash because the full-grown fruits can be stored for months. There is no universal agreement on how to handle the taxonomic treatment of the genus, with the number of species being listed varying from 13–30. There is uncertainty as to the ancestors of some species.

Most *Cucurbita* species are vines, grow several metres in length, have yellow or orange flowers, and may have tendrils. Non-vining cultivars of *C. pepo* have been developed. There are two types of flowers on a plant, the female flowers that produce the fruit, and the male flowers that produce pollen. Many North and Central American species are visited by specialist pollinators in the apid group, but other insects including honey bees also visit.

The *Cucurbita* genus is an important source of human food and the fruits are good sources of several nutrients such as vitamin A, vitamin C, dietary fibre, niacin, folic acid, and iron. In addition they are free of fat and cholesterol. The plants also contain the toxins cucurmosin and cucurbitacin. Medical uses of the plant include treating skin conditions and improving visual acuity.

Pests and Diseases

Cucurbita *species are used as food plants by the larvae of some Lepidoptera species, including Cabbage Moths,* Hypercompe indecisa, *and Turnip Moths.* Cucurbita *can be susceptible to the pest* Bemisia argentifolii *(silverleaf whitefly) as well as aphids (*Aphididae*), cucumber beetles (*Acalymma vittatum *and* Diabrotica undecimpunctata howardi*), squash bug (*Anasa tristis*), the squash vine borer (*Melittia cucurbitae*), and the twospotted spidermite (*Tetranychus urticae*). The squash bug causes major damage to plants because of its very toxic saliva. Cucurbits are susceptible to diseases such as bacterial wilt (*Erwinia tracheiphila*), anthracnose (*Colletotrichum *spp.), fusarium wilt (*Fusarium *spp.), phytophthora blight (*Phytophthora *spp. water molds), and powdery mildew (*Erysiphe *spp.). Defencive responses to viral, fungal, and bacterial leaf pathogens do not involve cucurbitacin.*

Species in the genus Cucurbita *are susceptible to some types of mosaic virus including: Cucumber mosaic virus (CMV), Papaya ringspot virus-cucurbit strain (PRSV), Squash mosaic virus (SqMV), Tobacco ringspot virus (TRSV), Watermelon mosaic virus (WMV), and Zucchini yellow mosaic virus (ZYMV). PRSV is the only one of these viruses that does not affect all cucurbits, which is the family Cucurbitaceae. SqMV and CMV are the most common viruses among cucurbits. Symptoms of these viruses show a high degree of similarity and laboratory investigation is often needed to differentiate which one is affecting plants.*

Lagenaria

Lagenaria is a genus of gourd-bearing vines from the family Cucurbitaceae, also known as the "Squash" family. It contains at least seven species, one of which is known as the Calabash (*Lagenaria siceraria*). Its species fruit can either be harvested young and used as a vegetable or harvested mature, dried, and used as a bottle or utensil.

Common names : bottle gourd, calabash (Eng.); kalbas (Afr.); moraka (North Sotho); segwana (Tswana); iselwa (Xhosa, Zulu)

Figure: *The bottle gourd is one of humankind's first domesticated plants, providing food, medicine and a wide variety of utensils and musical instruments.*

Description

Vigorous annual herb. Stems prostrate or climbing, angular, ribbed, thick, brittle, softly hairy, up to 5 m long, cut stems exude no sap. Leaves simple, up to 400 mm long and 400 mm broad, shortly and softly hairy, broadly egg-, kidney- or heart-shaped in outline, undivided, angular or faintly 3-7-lobed, lobes rounded, margins shallowly toothed, crushed leaves non-aromatic. Leaf stalks up to 300 mm long, thick, often hollow, densely hairy, with two small, lateral glands inserted at the leaf base. Tendrils split in two.

Flowers stalked (female flower stalks shorter than male), solitary, monoecious (male and female flowers on the same plant); petals 5, crisped, cream or white with darker veins, pale yellow at the base, obovate, up to 45 mm long, opening in the evenings, soon wilting. Fruit large, variable, up to 800 x 200 mm, subglobose to cylindrical, flask-shaped or globose with a constriction above the middle; fleshy, densely hairy to ultimately glabrous, indehiscent, green, maturing yellowish or pale brown, pulp drying out completely on ripening, leaving a thick, hard, hollow shell with almost nothing inside except the seeds. Seeds many, embedded in a spongy pulp, 7-20 mm long, compressed, with two flat facial ridges, in some variants rather irregular and rugose.

Distribution

It is generally accepted that *L. siceraria* (previously known as *L. vulgaris* Ser.) is indigenous to Africa and that it reached temperate and tropical areas in Asia and the Americas about 10 000 years ago, with human help or probably as a wild species whose fruits had floated across the seas. Fruits are known to float in the sea for many months without the seeds losing their viability. Independent domestications from wild populations are believed to have occurred in both the Old and New Worlds.

African and American land races (subsp. *siceraria*) are morphologically distinct from Asian land races (subsp. *asiatica*). It is uncertain if the seemingly spontaneous populations in Africa are truly wild. However, it is possible that Mary Wilkins-Ellert discovered an unusual free-living plant of *L. siceraria* in a remote region of southeastern Zimbabwe fairly recently (Decker-Walters *et al.* 2004). In southern Africa, *L. siceraria* has been collected in Namibia, Botswana, Swaziland

and the Limpopo, North-West and Gauteng provinces of South Africa. Most, if not all of these collections probably are from cultivated plants or plants that escaped from cultivation.

The calabash has been found growing mainly on alluvial sandy soil and red loam, on flat areas and moderate slopes, on rocky ridges, on river banks and in dry river beds, in disturbed areas and close to the road, often in the shade. It is common in riverine thickets, but also in woodland and savanna with *Acacia*, *Colophospermum mopane*, *Faidherbia albida* and *Phragmites*. It can also be found in cultivated lands with *Sorghum* and maize.

Derivation of the Name

The genus name *Lagenaria* comes from *lagena,* the Latin name for a Florence flask; referring to the fruit of *L. siceraria*. The species name *siceraria* probably also refers to the fruit which is useful when it is mature and dry (*siccus*).

The Cucurbitaceae consists of about 120 genera and 735 species that are cosmopolitan in mostly tropical and subtropical countries. Many species are cultivated and of economic importance as food plants such as cucumber, melon, pumpkin and watermelon. Members of this family are annual or perennial herbs or shrubs (only one species is a tree). The leaves are alternate and variable and tendrils are almost always present. The flowers are mostly unisexual and white or yellow; they occur on the same plant (monoecious) or on separate plants (dioecious). The fruit often is an indehiscent berry (soft-shelled) or gourd (hard-shelled) with one to many, often flattened seeds. There are about 18 genera and 75 species of this family in southern Africa.

The genus *Lagenaria* contains six species, probably all originally Old World and mainly African. Only two species are found in southern Africa; *L. siceraria* and *L. sphaerica*.

Ecology

In southern Africa the calabash flowers mainly in April and May and fruits mainly from April to July. In this region it grows from about 245-1 265 m above sea level, in areas with a rainfall of about 400-600 mm annually. Heiser (1997) observed that the bottle gourd flowers are pollinated at night by hawk-moths and other moth species; by day bumble bees, cucumber beetles and other insects are active. In the Americas hummingbirds are also attracted by the nectar that only the male flowers produce. He also observed that about 40 male and about 3 to 4 female flowers occur on a single plant.

Uses and Cultural Aspects

According to Watt & Breyer-Brandwijk (1962), human poisoning has been reported from eating the fruit, presumably of the bitter form. (As with other members of the Cucurbitaceae, e.g. *Cucumis africanus,* some plants of *L. siceraria* produce bitter, poisonous fruits and others non-bitter, edible fruits). Various medicinal uses of the leaves, fruit and seeds have been recorded from various countries, e.g. as a pectoral, an anthelmintic, a purgative and even as a headache remedy.

In southern Africa, the leaves are commonly eaten as a vegetable and are added fresh to maize porridge, or a relish is prepared from them, mixed with other plants. Dried leaves are stored for use in the lean season. The young shoots seem to be an important vegetable, unlike the young fruits that are considered by some to be a famine food. However, in some parts of southern Africa, the young, sweet and green fruit is a popular cooked vegetable, as in many parts of the world. Bitter fruits are not eaten and considered poisonous. Rood (1994) gives recipes for dishes with the young fruit of the calabash as the main ingredient. According to van Wyk & Gericke (2000), a form of small gourd with a coarse skin, is popular in southern Africa as a vegetable known as *maranka / maraka*, but this is actually a land race of the true squash, *Cucurbita pepo.*

Until about 25 years ago, South African farmers, particularly from the Ladismith District, Little Karoo, Western Cape, used to cultivate calabashes for the pipe-making industry. The necks of the young fruit were bent regularly in such a way that they grew to form the bowls of pipes. When the fruit was dry and ripe, these necks were sawn off, cleaned and exported to pipe makers overseas. It seems that these calabash pipes became very popular in foreign countries after the Anglo-Boer War, and around 1910 they were sold in the USA for up to 12 dollars. The US Dept. of Agriculture felt that these pipe bowls could be produced far more cheaply locally and in 1909 published a pamphlet detailing the cultivation of the South African pipe calabash and the preparation of these pipes.

Heiser wrote *The gourd book,* devoting more than half of it to *L. siceraria.* Numerous uses of the bottle gourd in various cultures are discussed and illustrated, even the use as penis sheaths in Papua New Guinea ! Kefalidou stated that bottle gourds were found all over the Mediterranean as early as the Bronze Age (3 rd and 2 nd millennia BC.). Containers crafted from gourds were in constant use in that region until recent times, as bottles for wine and water, dippers, salt containers, life-jackets (made of gourds tied together), and rattles.

Summit & Widess wrote a book on the rich diversity of musical instruments fashioned from bottle and other gourds in many cultures in various countries. Such instruments are played even today, also in the USA.

In rural southern Africa and elsewhere in developing countries, the calabash fruit is still widely used for various kinds of household containers and utensils. Non-bitter types are used to store water, milk and beer. These calabashes and products made from them are commonly sold as ornaments at roadside stalls and curio markets.

Pedro Paulo Salles from the University of São Paulo has provided some additional information regarding the use of bottle gourds in Brazil. Here they are known as cabaça, abóbora and porongo (or purunga). He notes that they are used for the transport of foods and liquids and for the construction of a range of musical instruments, some by native tribes. They are also used in cookery, especially in the state of Minas Gerais.

Citrullus

Citrullus is a small genus of desert vines, among which *Citrullus lanatus* (the watermelon) is an important crop.

Watermelon (*Citrullus lanatus* var. *lanatus*, family Cucurbitaceae) is a vine-like (scrambler and trailer) flowering plant originally from southern Africa. Its fruit, which is also called *watermelon*, is a special kind referred to by botanists as a pepo, a berry which has a thick rind (exocarp) and fleshy centre (mesocarp and endocarp). Pepos are derived from an inferior ovary, and are characteristic of the Cucurbitaceae. The watermelon fruit, loosely considered a type of melon – although not in the genus *Cucumis* – has a smooth exterior rind (usually green with dark green stripes or yellow spots) and a juicy, sweet interior flesh (usually deep red to pink, but sometimes orange, yellow, or white).

Varieties

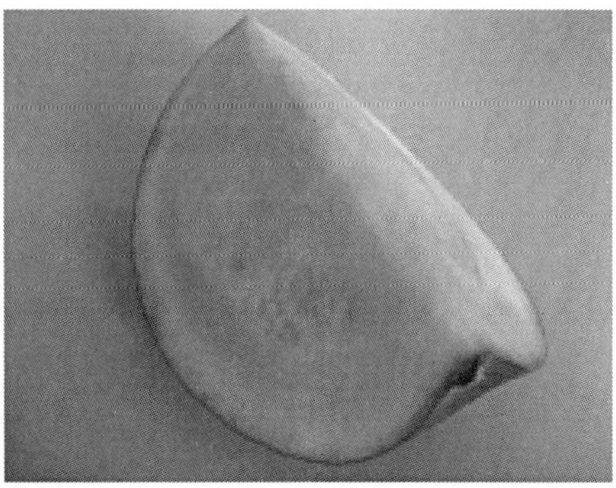

Figure: *Watermelon with yellow flesh*

The more than 1200 cultivars of watermelon range in weight from less than one to more than 200 pounds; the flesh can be red, orange, yellow or white.

- The 'Carolina Cross' produced the current world record watermelon weighing 262 pounds (119 kg). It has green skin, red flesh and commonly produces fruit between 65 and 150 pounds (29 and 68 kg). It takes about 90 days from planting to harvest.
- The 'Yellow Crimson' has a yellow-coloured flesh. It has been described as sweeter and more honey-flavoured than the more popular red-flesh watermelon.
- The 'Orangeglo' has a very sweet orange flesh, and is a large, oblong fruit weighing 9–14 kg (20–30 pounds). It has a light green rind with jagged dark green stripes. It takes about 90–100 days from planting to harvest.
- The 'Moon and Stars' variety was created in 1926. The rind is purple/black and has many small, yellow circles (stars) and one or two large, yellow circles (moon). The melon weighs 9–23 kg (20–50 pounds). The flesh is pink or red and has brown seeds. The foliage is also spotted. The time from planting to harvest is about 90 days.

Cucumis

Cucumis is a genus of twining, tendril-bearing plants in the Cucurbitaceae family which includes the cucumber (*Cucumis sativus*), true melons, the horned melon (*Cucumis metuliferus*), and the West Indian gherkin (*Cucumis anguria*). 30 species occur in Africa, and 25 occur in India, Southeast Asia, and Australia.

Figure: *Insect Pests and Their Management*

Pest Management

IPM tips for managing sucking pests:

- Site sanitation – weeds, infested plants, crop debris must be removed
- Use clean transplants – don't introduce the pests
- Monitor crops – early detection and population knowledge are a must
- Crop-free periods may work
- Avoid overlapping crops or kill pests in residue before planting again nearby
- Know the pests you have or are targeting
- Know the impacts of your insecticides on beneficials
- Understand resistance management and chemical groups
- Use chemical insecticides only when necessary and do not rely on them

IPM tips for managing chewing pests:

- Use resistant varieties
- Consider planting time of crops
- Identify and monitor the populations of both the pests and beneficials – including eggs, small larvae, and adults
- Understand available management options and effective bio-pesticides
- Understand conducive environmental conditions – for pests, beneficials, and biological insecticides
- Disrupt pest life cycles by targeting overwintering and survival sites
- Know your acceptable limits for damage and identify when you may need to spray

IPM for nematodes:

- Prevention
- Avoidance – Soil testing, rotations, use of fallow periods
- Monitoring
- Suppression

IPM for diseases:

- Understand the lifecycles and conducive environmental conditions for particular diseases

- Commit to good farm hygiene practices – clean up your farm and remove all weeds, crop debris, and volunteer hosts
- Monitor weather conditions (particularly temperature, humidity, and leaf wetness)
- Monitor crops regularly and use predictive models
- Use resistant varieties
- Reduce the pathogen levels by crop rotation
- Amend and manage soil to disadvantage fungi
- Avoid heavily infested blocks by testing soil prior to planting
- Exclude or eradicate the pathogen (quarantine; use of pathogen-tested seeds and propagated materials)
- Learn how to detect early disease symptoms on your crop
- Minimise ways in which the disease can spread on-farm – remove and destroy sick plants when symptoms first show
- Understand chemical resistance and rotate chemical groups

Beneficials include:

- Ladybird beetles
- Hover flies
- Lacewings
- Predatory bugs, thrips and mites
- Moth egg parasitoids, e.g. Trichogramma spp.
- Aphid parasitoids, e.g. Aphidius spp.
- Whitefly parasitoids, e.g Encarsia formosa
- Insect bacterial diseases – Bacillus thuringiensis (Bt)

4

Eggplant, Cowpea, Okra and Beans

Eggplant (*Solanum melongena*) is a species of nightshade commonly known in British English as aubergine and also known as melongene, garden egg, or guinea squash. It is known in South Asia, Southeast Asia and South Africa as brinjal. It bears a fruit of the same name (commonly either "eggplant" in American and Australian English or "aubergine" in British English) that is widely used in cooking, most notably as an important ingredient in dishes such as moussaka and ratatouille. As a member of the genus *Solanum*, it is related to both the tomato and the potato. It was originally domesticated in India and Bangladesh from the wild nightshade, the thorn or bitter apple, *S. incanum*.

Description

The eggplant is a delicate, tropical perennial often cultivated as a tender or half-hardy annual in temperate climates. It grows 40 to 150 cm (16 to 57 in) tall, with large, coarsely lobed leaves that are 10 to 20 cm (4–8 in) long and 5 to 10 cm (2–4 in) broad. Semiwild types can grow much larger, to 225 cm (7 ft) with large leaves over 30 cm (12 in) long and 15 cm (6 in) broad. The stem is often spiny. The flower is white to purple, with a five-lobed corolla and yellow stamens. The egg-shaped glossy black fruit has white flesh with a meaty texture. The cut surface of the flesh rapidly turns brown when the fruit is cut open. On wild plants, the fruit is less than 3 cm (1.2 in) in diameter, but very much larger in cultivated forms, reaching 30 cm (12 in) or more in length.

The fruit is botanically classified as a berry and contains numerous small, soft seeds which are edible, but have a bitter taste because they contain nicotinoid alkaloids (it is a close relative of tobacco).

Names and Etymology

Figure: *Closeup of an eggplant flower of a long-fruited Chinese variety in Hong Kong.*

Some 18th-century European cultivars were yellow or white and resembled goose or hen's eggs, hence the name "eggplant."

Many other names, some of which are superficially quite different, all derive ultimately from a Dravidian word, with modern reflexes in Kannada *Badanekayi*, Telugu *Vangakaya*, Malayalam *va_utina*, Tamil *va_utuGai*. This was borrowed into Sanskrit and Pali as *vâtiEgaGa*, *vâtigama*, which in turn was borrowed by Persian as *bâdingân* ÈÇÏäÌÇä, then by Arabic as *(al-)bâinjân* ÈÇÐäÌÇä.

The Arabic name is the common source of almost all European names for this plant, but through two distinct paths of transmission, with the *melongene* family coming through the eastern Mediterranean, and the *aubergine* family through the western Mediterranean.

In the eastern Mediterranean, Byzantine Greek borrowed *bâinjân* as ìåëéôæÜíα *melitzána*, influenced by Greek ìåëáíï- 'black'. That form came into medieval Latin as *melongena*, which was used in the botanical works of Tournefort and Linnaeus. Though *melongene* has become obsolete in the standard English, as has the French *melanjan*, it persists in the Caribbean English *melongene* or *meloongen*. The usual word in Italian remains *melanzana*.

Even the archaic English name *mad-apple* comes from the *melongena* family: in Italian, the word *melanzana* was reinterpreted in Italian as *mela insana*, and translated into English as *mad apple*.

In the western Mediterranean, *(al)-bâinjân* became Spanish *berenjena*, Catalan as *albergínia*, and Portuguese *beringela*. The Catalan form was borrowed by French as *aubergine*, which was then borrowed into British English.

In Eastern Slavic languages, such as Russian and Ukrainian, the word *baklazhan* is used, while Turkish has *patlýcan*.

In South Asian, South African, Malaysian, Singaporean, and West Indian English, the fruit is called *brinjal*, from the Portuguese. The Indic name *baingan* or *baigan* is also sometimes used in South Asian English.

In Kiswahili, it is called *biringanya*.

Cultivated Varieties

Figure: *Three varieties of eggplant.*

Different varieties of the plant produce fruit of different size, shape, and colour, though typically purple. The most widely cultivated varieties (cultivars) in Europe and North America today are elongated ovoid, 12–25 cm long (4½ to 9 in) and 6–9 cm broad (2 to 4 in) in a dark purple skin.

A much wider range of shapes, sizes and colours is grown in India and elsewhere in Asia. Larger varieties weighing up to a kilogram (2.2 pounds) grow in the region between the Ganges and Yamuna rivers, while smaller varieties are found elsewhere. Colours vary from white to yellow or green, as well as reddish-purple and dark purple. Some cultivars have a colour gradient, from white at the stem to bright pink to deep purple or even black. Green or purple cultivars in white striping also exist. Chinese varieties are commonly shaped like a narrower, slightly pendulous cucumber, and are sometimes called Japanese eggplants in North America.

Oval or elongated oval-shaped and black-skinned cultivars include 'Harris Special Hibush', 'Burpee Hybrid', 'Black Magic', 'Classic', 'Dusky', and 'Black Beauty'. Slim cultivars in purple-black skin include 'Little Fingers', 'Ichiban', 'Pingtung Long', and 'Tycoon'; in green skin, 'Louisiana Long Green' and 'Thai (Long) Green'; in white skin, 'Dourga'. Traditional, white-skinned, egg-shaped cultivars include 'Casper' and 'Easter Egg'. Bicoloured cultivars with colour gradient include 'Rosa Bianca', 'Violetta di Firenze', 'Bianca Smufata di Rosa' (heirloom), and 'Prosperosa' (heirloom). Bicoloured cultivars with striping include 'Listada de Gandia' and 'Udumalapet'. In some parts of India, miniature varieties (most commonly called *vengan*) are popular. A particular variety of green brinjal known as *Matti gulla* is grown in Matti, a village of the Udupi district in Karnataka state.

Eggplant Pests and Diseases

Eggplant is a commonly grown warm-season vegetable most noted for its great flavour in addition to its dark violet colour and egg shape. Several other varieties can be grown in the home garden as well. They consist of various colours and sizes, all of which can add unique flavour to many dishes or as stand-alone side dishes. Eggplant problems and eggplant pests can occur from time to time when growing eggplant; however, with the proper care, these can usually be prevented.

Growing Eggplant

Eggplants are very cold sensitive and should not be placed into the garden too early. Wait until the soil has sufficiently warmed and all threat of frost has ceased. These plants require full sun and well-drained soil amended with organic matter.

When growing eggplants, you'll want to space them about a foot or two apart, as they can become rather large. Since eggplants are susceptible to many pests and diseases, the use of collars or row covers on young plants may be necessary in reducing these common eggplant problems.

Dealing with Eggplant Pests

Lace bugs and flea beetles are common eggplant bugs. Other eggplant bugs that affect these plants include:

Tomato Hornworms

The five-spotted hawkmoth (*Manduca quinquemaculata*) is a brown and gray hawk moth of the Sphingidae family. The caterpillar, often referred to as the tomato hornworm, can be a major pest in gardens. Tomato hornworms are closely related to (and sometimes confused with) the tobacco hornworm (*Manduca sexta*). This confusion arises because caterpillars of both species feed on the foliage of various plants from the family Solanaceae, so either species can be found on tobacco or tomato leaves, and the plant on which the caterpillar is found does not indicate its species. The larvae of these species can be distinguished by their lateral markings: tomato hornworms have seven V-shaped markings while tobacco hornworms have seven diagonal lines. Furthermore, the caterpillars can be distinguished from the larval stage onwards by the colour of the horns on their back ends: *M. quinquemaculata* caterpillars have black horns, while *M. sexta* caterpillars have red horns. The moths can be distinguished by the number of spots on their abdomens, with *M. quinquemaculata* having, as its name suggests, five.

The best way to deal with eggplant bugs is by using collars and row covers until the plants are large enough to withstand attacks, at which time insecticidal soaps can be used to alleviate pest problems.

To prevent eggplant bugs, it may also help to keep weeds and other debris to a minimum and rotate crops every other year or so. Introducing natural predators, such as ladybugs, often helps minimize eggplant problems associated with aphids.

How to Get Rid of Lace Bug Pests

A reddish-orange colour on the undersides of the leaves on your trees and shrubs is a good sign that you are dealing with lace bugs. These tiny insects can ruin the appearance of your landscape once they begin feeding on your plants. Here are some tips on how to get rid of lace bug pests.

Lace bugs are tiny insects that grow no more than one-eighth inch long. Small, clear cells cover their wings and thorax, giving them their lacy appearance. They feed by sucking the sap from the foliage of trees and shrubs, leaving them looking mottled, stippled and discoloured.

Dealing with lace bugs can be annoying but the good news is that with effective lace bug treatment, you can rid them of the garden.

Natural Control of Lace Bugs

There are dozens of species of lace bugs, and each one feeds on only one species of plant. For example, a walnut lace bug won't feed on an azalea, and a willow lace bug won't feed on a sycamore. Therefore, planting a wide variety of species in the landscape prevents the insect from spreading.

Another method of natural control of lace bugs is to take advantage of the fact that lace bugs are more likely to feed on plants in hot, dry and sunny environments. Work compost into the soil and mulch around the plants to keep the soil evenly moist. Also, provide afternoon shade whenever possible.

Lace Bug Treatment with Insecticides

A number of beneficial insects help keep lace bugs under control, including:

- jumping spiders
- assassin bugs
- lacewing larvae
- pirate bugs
- lady beetles
- predaceous mites

Avoid using broad-spectrum insecticides that destroy the lace bug's predators. Once they are gone, the plant has no natural defence against lace bugs, and you might develop a spider mite problem.

Instead, use insecticidal soap, neem oil or narrow-range oil. Spray the plant with these insecticides at two-week intervals. The damage won't disappear, but you won't have any new damage.

Eggplant Diseases in the Garden

There are several eggplant diseases that affect these crops. Some of the most common include blossom end rot, wilt diseases, and various types of blight. Many of these eggplant diseases can be eliminated or prevented by practicing crop rotation, reducing weed growth, and providing adequate spacing and uniform watering.

- Blossom end rot, as found in tomatoes, is caused from fungus due to overwatering and affects ripe fruit. Round, leathery, sunken spots appear on fruit ends with the affected fruit eventually dropping from the plant.
- Bacterial wilt can cause plants to suddenly droop, from the bottom to the top, turning yellow. Affected plants eventually wither up and die.
- Verticillium wilt is similar to bacterial wilt but is caused from soil-borne fungal infections. Plants may become stunted, turn yellow, and wilt.
- Southern blight is also caused by fungus and plants exhibit softening of the crown and root tissues. Mold may also be seen on the stems and surrounding soil.

- Phomopsis blight usually affects fruits of eggplant, which begin as sunken spots that eventually enlarge and become soft and spongy. Leaves and stems, especially seedlings, may develop gray or brown spots first.
- Phytophthora blight, which also affects peppers, can quickly destroy eggplants. Plants will get dark streaks prior to collapsing and dying.

Cowpea

The cowpea (*Vigna unguiculata*) is one of several species of the widely cultivated genus *Vigna*. Four subspecies are recognised, of which three are cultivated (more exist, including *textilis*, *pubescens*, and *sinensis*):

Cowpeas are one of the most important food legume crops in the semiarid tropics covering Asia, Africa, southern Europe and Central and South America. A drought-tolerant and warm-weather crop, cowpeas are well-adapted to the drier regions of the tropics, where other food legumes do not perform well. It also has the useful ability to fix atmospheric nitrogen through its root nodules, and it grows well in poor soils with more than 85% sand and with less than 0.2% organic matter and low levels of phosphorus. In addition, it is shade tolerant, so is compatible as an intercrop with maize, millet, sorghum, sugarcane, and cotton. This makes cowpeas an important component of traditional intercropping systems, especially in the complex and elegant subsistence farming systems of the dry savannas in sub-Saharan Africa. In these systems the haulm (dried stalks) of cowpea is a valuable by-product, used as animal feed.

Figure: *Cowpea beans*

Research in Ghana found that selecting early generations of cowpea crops to increase yield is not an effective strategy. Francis Padi from the Savannah Agricultural Research Institute in Tamale, Ghana, writing in *Crop Science*, suggests other methods such as bulk breeding are more efficient in developing high-yield varieties.

According to the USDA food database, the leaves of the cowpea plant have the highest percentage of calories from protein among vegetarian foods.

Taxonomy and Etymology

Vigna unguiculata is a member of the vigna (peas or beans) genus. Unigiculata is Latin for "with a small claw", which reflects the small stalks on the flower petals. All cultivated cowpeas are found within the universally accepted V. unguiculata subspecies unguiculata classification, which is then commonly divided into four cultivar groups: Unguiculata, Biflora, Sesquipedalis, and Textilis. Some of the more well known common names for cultivated cowpeas include black-eye pea, southern pea, yardlong bean, catjang and Crowder Pea. The classification of the wild relatives within V. unguiculata is more complicated, with over 20 different names having been used and between 3 and 10 subgroups described. The original subgroups of stenophylla, dekindtiana and tenuis appear to be common in all taxonomic treatments, with the earlier described variations pubescens and protractor being raised to sub species level by a 1993 charactisation.

Common Names of Vigna Unguiculata Cultivar Groups

Group	*Common name*
Unguiculata	crowder-pea, southern pea, black-eyed pea
Biflora	catjang, sow-pea
Sesquipedalis	yardlong bean, asparagus bean, Chinese long-bean
Textilis	

The first written reference using cowpea appeared in 1798 in the United States. The name was most likely acquired due to their use as a fodder crop for cows. The common name of black-eyed pea, used for the unguiculata cultivar group, describes the presence of a distinctive black spot at the hilum of the seed. Black-eyed peas were first introduced to the southern states in the United States and some early varieties had peas squashed closely together in their pods, leading to the other common names of southern pea and crowder-pea. Sesquipedalis in Latin means "foot and a half long", and this subspecies which arrived in the United States via Asia is characterised by unusually long pods, leading to the common names of yardlong bean, asparagus bean and Chinese long-bean. In West Africa the plant is named niebe, wake or ewa.

Description

There is a large morphological diversity found within the crop, and the growth conditions and grower preferences for each variety vary from region to region.

History

Although there is no archaeological evidence for early cowpea cultivation the centre of diversity of the cultivated cowpea is West Africa, leading to the current consensus that this is the likely centre of origin and place of early domestication. In 2300 BC the cowpea is believed to have made its way into South East Asia where secondary domestication events may have occurred. The first written references to the cowpea were in 300BC and they probably reached Central and North America during the slave trade through the 17th to early 19th centuries.

Cultivation

Cowpeas are grown mostly for their edible beans, although the leaves, fresh peas and fresh pea pods can also be consumed, meaning the cowpea can be used as a food source before the dried peas are harvested. Cowpeas thrive in poor dry conditions, growing well in soils up to 85% sand. This makes them a particularly important crop in arid, semi-desert regions where not many other crops will grow. As well as an important source of food for humans in poor arid regions the crop can also be used as feed for livestock. This predominately occurs in India, where the stock is fed cowpea as forage or fodder. The nitrogen fixing ability means that as well as functioning as a sole-crop, the cowpea can be effectively intercropped with sorghum, millet, maize, cassava or cotton.

Culinary Use

In Tamil Nadu, India, between the Tamil months of Maasi (February) and Panguni (March), a cake-like dish called kozhukattai (steamed sweet dumplings - also called adai in Kerala) is prepared with cooked and mashed cowpeas mixed with jaggery, ghee, and other ingredients. Thatta payir in sambar and pulikkuzhambu (spicy semisolid gravy in tamarind paste) is a popular dish in Tamil Nadu.

In Sri Lanka, cowpeas are cooked in many different ways, one of which is with coconut milk.

In Turkey, cowpeas can be lightly boiled, covered with olive oil, salt, thyme, and garlic sauce, and eaten as an appetizer. Also, they are cooked with garlic and tomatoes. And they can be eaten in bean salad.

Nutrition and Health

Cowpeas provide a rich source of proteins and calories, as well as minerals and vitamins. A cowpea seed can consist of 25% protein and is low in anti-nutritional factors. This diet complements the mainly cereal diet in countries that grow cowpeas as a major food crop.

Production and Consumption

Most cowpeas are grown on the African continent, particularly in Nigeria and Niger which account for 72% of world cowpea production. The Sahel region also contains other major producers such as Burkina Faso, Ghana, Senegal and Mali. Niger is the main exporter of cowpeas and Nigeria the main importer. Exact figures for cowpea production are hard to come up with as it is not a major export crop. A 1997 estimate suggests that cowpeas are cultivated on 12.5 million hectares and have a worldwide production of 3 million tonnes. While they play a key role in subsistence farming and livestock fodder, the cowpea is also seen as a major cash crop by Central and West African farmers, with an estimated 200 million people consuming cowpea on a daily basis.

According to the Food and Agriculture Organisation of the United Nations (FAO), as of 2012, the average cowpea yield in Western Africa was an estimated 483 kg/ha, which is still 50% below the estimated potential production yield. In some tradition cropping methods the yield can be as low as 100 kg/ha.

Outside Africa, the major production areas are Asia, Central America, and South America. Brazil is the world's second-leading producer of cowpea seed, producing 600,000 tonnes annually. The amount of protein content of cowpea's leafy parts consumed annually in Africa and Asia is equivalent to 5 million tonnes of dry cowpea seeds, representing as much as 30% of the total food legume production in the lowland tropics.

Okra

Okra known in many English-speaking countries as ladies' fingers, bhindi, bamia, or gumbo, is a flowering plant in the mallow family. It is valued for its edible green seed pods. The geographical origin of okra is disputed, with supporters of South Asian, Ethiopian and West African origins. The plant is cultivated in tropical, subtropical and warm temperate regions around the world.

Structure and Physiology

The species is an annual or perennial, growing to 2 m tall. It is related to such species as cotton, cocoa, and hibiscus. The leaves are

10–20 cm long and broad, palmately lobed with 5–7 lobes. The flowers are 4–8 cm in diameter, with five white to yellow petals, often with a red or purple spot at the base of each petal. The fruit is a capsule up to 18 cm long, containing numerous seeds.

Abelmoschus esculentus *is cultivated throughout the tropical and warm temperate regions of the world for its fibrous fruits or pods containing round, white seeds. It is among the most heat- and drought-tolerant vegetable species in the world and will tolerate soils with heavy clay and intermittent moisture but frost can damage the pods.*

In cultivation, the seeds are soaked overnight prior to planting to a depth of 1–2 cm. Germination occurs between six days (soaked seeds) and three weeks. Seedlings require ample water. The seed pods rapidly become fibrous and woody, and, to be edible, must be harvested within a week of the fruit having been pollinated. The fruits are harvested when immature and eaten as a vegetable.

Origin and Distribution

Okra is an allopolyploid of uncertain parentage (proposed parents include *Abelmoschus ficulneus*, *A. tuberculatus* and a reported "diploid" form of okra). Truly wild (as opposed to naturalised) populations are not known with certainty and the species may be a cultigen.

The geographical origin of okra is disputed, with supporters of South Asian, Ethiopian and West African origins. Supporters of a South Asian origin point to the presence of its proposed parents in that region. Supporters of a West African origin point to the greater diversity of okra in that region.

Figure: *Lady's fingers (Okra) in Chennai, India.*

Figure: *Okra sliced.*

The Egyptians and Moors of the 12th and 13th centuries used the Arabic word for the plant, *bamya*, suggesting it had come from the east. The plant may have entered southwest Asia across the Red Sea or the Bab-el-Mandeb strait to the Arabian Peninsula, rather than north across the Sahara, or from India. One of the earliest accounts is by a Spanish Moor who visited Egypt in 1216, who described the plant under cultivation by the locals who ate the tender, young pods with meal.

From Arabia, the plant spread around the shores of the Mediterranean Sea and eastward. The plant was introduced to the Americas by ships plying the Atlantic slave trade by 1658, when its presence was recorded in Brazil. It was further documented in Suriname in 1686. In compound farms in the rainforest of southeastern Nigeria (Okafor and Fernandes, 1986), farmers have developed a multi-crop system that provides a diversified and continuous production of food, combining species with different maturity periods such as yams, cassava, cocoyams, bananas, plantain, maize, okra, pumpkin, melon, leafy vegetables and a variety of trees and shrubs, 60 of which provide food products. This ensures a balanced diet but also reduces the need for storage in an area where post-harvest losses are high.

Okra may have been introduced to southeastern North America in the early 18th century. It was being grown as far north as Philadelphia by 1748. Thomas Jefferson noted it was well established in Virginia by 1781. It was commonplace throughout the Southern United States by 1800, and the first mention of different cultivars was in 1806.

Okra has few noticeable pests. Pods curled or with wart-like protrusions commonly indicate earlier feeding damage by stink bugs and/or leaffooted bugs. Corn earworms also infest pods occasionally.

Damage to the foliage is normally unimportant, unless Japanese beetle populations are unusually high.

Key to Okra Pests

Insects that feed externally on buds, foliage, or pods

1. Chewing insects that leave holes in plants
   a. *Corn earworm* - Early instar cream coloured or yellowish-green with few markings; later instar green, reddish, or brown with pale longitudinal stripes and scattered black spots; moderately hairy; up to 44 mm long; three pairs of legs, five pairs of prolegs; eats holes in pods.
   b. *Japanese beetle* - Hard-bodied, shiny, metallic green beetle with coppery brown wing covers which meet in a straight line down the centre of the back; body about 13 mm long; skeletonizes leaves; may also eat pods
2. Pests that cause discolouration or distortion of plant
   a. *Aphids* - Soft-bodied, pear-shaped insects with a pair of dark cornicles and a cauda protruding from the abdomen; may be winged or wingless - wingless forms most common; feed in colonies; excrete honeydew on which sooty mold grows.

1. Green peach aphid - Pale yellow to green wingless adult up to 2.4 mm long; winged adult with dark dorsal blotch on yellowish-green abdomen; cornicles over twice as long as cauda and slightly swollen toward tip; yellow-green nymph with three dark lines on abdomen.
2. *Melon aphid* - Yellow or green in host, dry summers; pale to dark green in cool seasons; body up to 2 mm long; cornicles slightly longer than or twice as long as cauda; nymph similar in colour to adult.

   b. *Leaffooted bugs and Stink bugs* - Bugs up to 19 mm long; adult green or brown and shield-shaped (stink bug), or brown with enlarged, flattened hind legs (leaffooted bug); stink bug nymph pale green or green with orange and black markings; leaffooted bug nymph red; pierce buds, pods, and seeds causing pods to be deformed and seeds to be shriveled; wart-like growths appear on pods.

   B. Insects that feed inside the plant, e.g., within the stalk or leaf mesophyll

1. European corn borer - Pale yellow or pinkish caterpillar up to 26 mm long with several rows of tiny, darkly coloured spots;

three pairs of legs near the head; five pairs of prolegs; leaves entrance hole fringed with silk and frass in stalk; weakens stem and interferes with pod development.

2. Vegetable leafminer - Colourless to bright yellow maggot with black mouthparts; body up to 3 mm long; pointed head; S-shaped mines in leaves, often enlarged at one end.

Beans

Bean is a common name for large plant seeds used for human food or animal feed of several genera of the family Fabaceae (alternately Leguminosae).

Cultivation

Figure: *Field beans (broad beans,* Vicia faba, *ready for harvest*

Unlike the closely related pea, beans are a summer crop that needs warm temperatures to grow. Maturity is typically 55–60 days from planting to harvest. As the bean pods mature, they turn yellow and dry up, and the beans inside change from green to their mature colour. As a vine, bean plants need external support, which may be provided in the form of special "bean cages" or poles. Native Americans customarily grew them along with corn and squash (the so-called Three Sisters), with the tall cornstalks acting as support for the beans.

In more recent times, the so-called "bush bean" has been developed which does not require support and has all its pods develop simultaneously (as opposed to pole beans which develop gradually). This makes the bush bean more practical for commercial production.

History

Beans are one of the longest-cultivated plants. Broad beans, also called fava beans, in their wild state the size of a small fingernail, were gathered in Afghanistan and the Himalayan foothills. In a form improved from naturally occurring types, they were grown in Thailand since the early seventh millennium BCE, predating ceramics. They were deposited with the dead in ancient Egypt. Not until the second millennium BCE did cultivated, large-seeded broad beans appear in the Aegean, Iberia and transalpine Europe. In the Iliad (late-8th century) is a passing mention of beans and chickpeas cast on the threshing floor.

Beans were an important source of protein throughout Old and New World history, and still are today.

The oldest-known domesticated beans in the Americas were found in Guitarrero Cave, an archaeological site in Peru, and dated to around the second millennium BCE.

Most of the kinds commonly eaten fresh or dried, those of the genus Phaseolus, come originally from the Americas, being first seen by a European when Christopher Columbus, during his exploration of what may have been the Bahamas, found them growing in fields. Five kinds of Phaseolus beans were domesticated by pre-Columbian peoples: common beans (Phaseolus vulgaris) grown from Chile to the northern part of what is now the United States, and lima and sieva beans (Phaseolus lunatus), as well as the less widely distributed teparies (Phaseolus acutifolius), scarlet runner beans (Phaseolus coccineus) and polyanthus beans (Phaseolus polyanthus) One especially famous use of beans by pre-Columbian people as far north as the Atlantic seaboard is the "Three Sisters" method of companion plant cultivation:

In the New World, many tribes would grow beans together with maize (corn), and squash. The corn would not be planted in rows as is done by European agriculture, but in a checkerboard/hex fashion across a field, in separate patches of one to six stalks each.

Beans would be planted around the base of the developing stalks, and would vine their way up as the stalks grew. All American beans at that time were vine plants, "bush beans" having been bred only more recently. The cornstalks would work as a trellis for the beans, and the beans would provide much-needed nitrogen for the corn.

Squash would be planted in the spaces between the patches of corn in the field. They would be provided slight shelter from the sun by the corn, would shade the soil and reduce evaporation, and would deter many animals from attacking the corn and beans because their coarse, hairy vines and broad, stiff leaves are difficult or uncomfortable for animals such as deer and raccoons to walk through, crows to land on, etc.

Dry beans come from both Old World varieties of broad beans (fava beans) and New World varieties (kidney, black, cranberry, pinto, navy/haricot).

Beans are a heliotropic plant, meaning that the leaves tilt throughout the day to face the sun. At night, they go into a folded "sleep" position.

Peas

The pea is most commonly the small spherical seed or the seed-pod of the pod fruit *Pisum sativum*. Each pod contains several peas. Peapods are botanically a fruit, since they contain seeds developed from the ovary of a (pea) flower. The name is also used to describe other edible seeds from the Fabaceae such as the pigeon pea (*Cajanus cajan*), the cowpea (*Vigna unguiculata*), and the seeds from several species of *Lathyrus*.

P. sativum is an annual plant, with a life cycle of one year. It is a cool season crop grown in many parts of the world; planting can take place from winter to early summer depending on location. The average pea weighs between 0.1 and 0.36 grams. The immature peas (and in snow peas the tender pod as well) are used as a vegetable, fresh, frozen or canned; varieties of the species typically called field peas are grown to produce dry peas like the split pea shelled from the matured pod. These are the basis of pease porridge and pea soup, staples of medieval cuisine; in Europe, consuming fresh immature green peas was an innovation of Early Modern cuisine.

The wild pea is restricted to the Mediterranean basin and the Near East. The earliest archaeological finds of peas date from the late neolithic era of current Greece, Syria, Turkey and Jordan. In Egypt, early finds date from *ca.* 4800–4400 BC in the Nile delta area, and from *ca.* 3800–3600 BC in Upper Egypt. The pea was also present in Georgia in the 5th millennium BC. Farther east, the finds are younger. Peas were present in Afghanistan *ca.* 2000 BC, in Harappa, Pakistan, and in northwest India in 2250–1750 BC. In the second half of the 2nd millennium BC, this pulse crop appears in the Gangetic basin and southern India.

A pea is a most commonly green, occasionally purple or golden yellow, pod-shaped vegetable, widely grown as a cool season vegetable crop. The seeds may be planted as soon as the soil temperature reaches 10 °C (50 °F), with the plants growing best at temperatures of 13 to 18 °C (55 to 64 °F). They do not thrive in the summer heat of warmer temperate and lowland tropical climates, but do grow well in cooler, high altitude, tropical areas. Many cultivars reach maturity about 60 days after planting.

Pests and Diseases

A variety of diseases affect peas through a number of pathogens, including insects, viruses, bacteria and fungi. In particular, virus disease of peas has worldwide economic importance.

Additionally, insects such as the pea leaf weevil (*Sitona lineatus*) can damage peas and other pod fruits. The pea leaf weevil is native to Europe, but has spread to other places such as Alberta, Canada. They are about 3.5 millimetres (0.14 in)—5.5 millimetres (0.22 in) long and are distinguishable by three light-coloured stripes running length-wise down the thorax. The weevil larvae feed on the root nodules of pea plants, which are essential to the plants' supply of nitrogen, and thus diminish leaf and stem growth. Adult weevils feed on the leaves and create a notched, "c-shaped" appearance on the outside of the leaves.

Pests of Beans and Peas

The pests of beans, southern peas, and English peas are a diverse group. Mites and beetles are usually the most common pests of beans.

Aphids frequently infest English peas, and stink bugs and leaffooted bugs are nuisances of southern peas. Some aphids transmit virus diseases.

Key to Pests of Beans and Peas

A. Pests that feed primarily on the foliage

1. Insects that mine or eat holes in the foliage

   a. *Bean leaf beetle* - These reddish to yellowish-brown beetles are 5 to 6 mm long and often have three black spots on each wing cover. They have black margins and a black triangle on the front portion of the wings. Bean leaf beetles consume mostly young leaves although the outer wall of pods are sometimes attacked when vegetative growth ends.

   b. *Mexican bean beetle* - Copper red, dome-shaped Mexican bean beetles are 6 to 8.5 mm long; each wing cover has eight small black spots that form three rows across the body when wings are at rest. These beetles skeletonize leaves which become lace-like and eventually turn brown from their feeding.

   c. *Vegetable leafminer* - Colourless to bright yellow, vegetable leafminers are maggots which grow up to 3 mm long. The head is pointed. These maggots make serpentine mines, slightly enlarged at the new end.

2. Sap-sucking pests that cause discolouration, deformation, or abscission

   a. *Aphids* - Aphids are soft-bodied, pear-shaped insects with a pair of dark cornicles (tailpipe-like appendages) and a cauda (tail) protruding from the abdomen; they may be winged or wingless - wingless forms most common. Aphids feed in colonies and cause discolouration, curling, and deformation of foliage. They often transmit virus diseases, and they excrete honeydew on which sooty mold grows.

      1. *Bean aphid* - Dark green to black, bean aphids have white appendages. They grow to 2.6 mm, and the cornicles are about the same length as the cauda. Nymphs are green; mature nymphs have five to seven pairs of white spots on abdomen.

      2. *Cowpea aphid* - Cowpea aphids are black with white appendages and up to 2.5 mm long. The cornicles are barely longer than the cauda. Nymphs are pale green to gray with a powdery coating.

3. *Melon aphid* - Yellow to green, melon aphids have dark cornicles and cauda. They feed in colonies and wingless forms are most common.
4. *Potato aphid* - Pink, mottled or light green, potato aphids have a dark stripe. Adults are up to 3.5 mm long, and the cornicles are slender and about twice as long as the cauda.<=”” ol=””>
5. *Potato leafhopper* - Spindle-shaped and up to 3 mm long, potato leafhoppers are green with yellowish to dark green spots. They usually jump instead of fly when disturbed. Potato leafhoppers extract sap from the undersides of leaves causing them to crinkle and curl downward. Infested plants become yellow or bronzed and dwarfed.
6. *Thrips* - Spindle-shaped and 1.2 mm or less in length, thrips are yellow, amber, brown, or black. Adults have two pairs of wings which are fringed and have brown crossbands. Immature thrips are white or yellow with red eyes. Thrips are pests during hot, dry weather. They cause whitish flecks or streaks on leaves and blossoms, and they deposit black specks of excrement.
7. *Twospotted spider mite* - Tiny (almost microscopic) pale to dark green, twospotted spider mites have two or four darkly coloured spots. Adults and nymphs have eight legs. Larvae have six legs. Females are oval and 0.3 to 0.5 mm long. Males are somewhat diamond shaped. Infested foliage becomes silvery because of pale yellow stipples. Leaves eventually become pale and die. Silken webs are spun on the underside of leaves as the mites feed.

B. Insects that feed on pods

8. *Corn earworm* - Early instars of corn earworms are cream coloured or yellowish green with few markings. Later instars are green, reddish or brown with pale longitudinal stripes and scattered black spots. Corn earworms are moderately hairy and grow to 44 mm long. They have three pairs of legs and five pairs of prolegs. They attack beans in fall and eat holes in pods.
9. *European corn borer* - These caterpillars are grayish pink with a dark head and rows of small brown doughnut-shaped spots on the back. European corn borers grow to about 26 mm long and bore into the pods.

10. *Cowpea curculio adult and larva* - Adults are black humpbacked weevils 6 to 7 mm long. Larvae are pale yellow and have brown heads. Larvae are legless and grow 6 to 7 mm long. Curculios leave feeding scars - small holes in pods and peas; larvae feed inside green seeds.

11. *Stink bugs* - Adults are green or brown shield-shaped insects up to 19 mm long; nymphs are pale green or green with orange and black markings. Stink bugs pierce buds, pods, and seeds and cause buds to be malformed and plants weakened.

C. Insects which damage seeds and roots and bore in stems

12. *Bean leaf beetle larva* - The whitish larva (up to 10 mm long) are dark at both ends and have three pairs of prolegs near the head. They are minor pests of bean roots.

13. *Lesser cornstalk borer* - These slender, bluish-green caterpillars are up to 19 mm long and have brown rings around the body, three pairs of legs near the head, and five pairs of prolegs on the abdomen. Young larva bore into stems and sometimes disrupt the growing point.

14. *Limabean vine borer* - Gray when young, these caterpillars later become bluish-green and sparsely covered with long yellowish hairs. Linnaean vine borer caterpillars grow up to 25 mm long and have three pairs of legs near the head and five pairs of prolegs on the abdomen. They move from the leaves into stems, usually near nodes where they cause galls up to 70 mm long and 20 mm around to develop. Short, loose, silky frass tubes are connected to entrance holes.

15. *Seedcorn maggot* - White to yellow-white maggots up to 7 mm long feed on seed contents causing poor germination and tall spindly seedlings. Seedcorn maggots have no legs. The head is pointed.

5

Potato and Sweet Potato

The potato is a starchy, tuberous crop from the perennial nightshade *Solanum tuberosum* L. The word "potato" may refer to the plant itself in addition to the edible tuber. In the Andes, where the species is indigenous, there are some other closely related cultivated potato species. Potatoes were introduced outside the Andes region four centuries ago, and have become an integral part of much of the world's food supply. It is the world's fourth-largest food crop, following maize, wheat and rice.

Wild potato species occur throughout the Americas, from the United States to southern Chile. The potato was originally believed to have been domesticated independently in multiple locations, but later genetic testing of the wide variety of cultivars and wild species proved a single origin for potatoes in the area of present-day southern Peru and extreme northwestern Bolivia (from a species in the *Solanum brevicaule* complex), where they were domesticated 7,000–10,000 years ago. Following centuries of selective breeding, there are now over a thousand different types of potatoes. Over 99% of the presently cultivated potatoes worldwide arc descended from varieties that originated in the lowlands of south-central Chile, which have displaced formerly popular varieties from the Andean highlands.

The annual diet of an average person in the first decade of the 21st century included about 33 kg (73 lb) of potato. However, the local importance of potato is extremely variable and rapidly changing. It remains an essential crop in Europe (especially eastern and central Europe), where per capita production is still the highest in the world, but the most rapid expansion over the past few decades has occurred in southern and eastern Asia. China is now the world's largest potato-producing country, and nearly a third of the world's potatoes are harvested in China and India.

Genetics

There are about 5,000 potato varieties worldwide. Three thousand of them are found in the Andes alone, mainly in Peru, Bolivia, Ecuador, Chile, and Colombia. They belong to eight or nine species, depending on the taxonomic school. Apart from the 5,000 cultivated varieties, there are about 200 wild species and subspecies, many of which can be cross-bred with cultivated varieties, which has been done repeatedly to transfer resistances to certain pests and diseases from the gene pool of wild species to the gene pool of cultivated potato species. Genetically modified varieties have met public resistance in the United States and in the European Union.

The major species grown worldwide is Solanum tuberosum *(a tetraploid with 48 chromosomes), and modern varieties of this species are the most widely cultivated. There are also four diploid species (with 24 chromosomes):* S. stenotomum, S. phureja, S. goniocalyx, *and* S. ajanhuiri. *There are two triploid species (with 36 chromosomes):* S. chaucha *and* S. juzcpczukii. *There is one pentaploid cultivated species (with 60 chromosomes):* S. curtilobum. There are two major subspecies of Solanum tuberosum: andigena, or Andean; and tuberosum, or Chilean. The Andean potato is adapted to the short-day conditions prevalent in the mountainous equatorial and tropical regions where it originated. The Chilean potato, native to the Chiloé Archipelago, is adapted to the long-day conditions prevalent in the higher latitude region of southern Chile.

The International Potato Centre, based in Lima, Peru, holds an ISO-accredited collection of potato germplasm. The international Potato Genome Sequencing Consortium announced in 2009 that they had achieved a draft sequence of the potato genome. The potato genome contains 12 chromosomes and 860 million base pairs making it a medium-sized plant genome. More than 99 percent of all current varieties of potatoes currently grown are direct descendants of a subspecies that once grew in the lowlands of south-central Chile. Nonetheless, genetic testing of the wide variety of cultivars and wild species affirms that all potato subspecies derive from a single origin in the area of present-day southern Peru and extreme northwestern Bolivia (from a species in the Solanum brevicaule complex).

Most modern potatoes grown in North America arrived through European settlement and not independently from the South American sources. However, at least one wild potato species, Solanum fendleri, *is found as far north as Texas and used in breeding for resistance to a nematode species that attacks cultivated potatoes. A secondary centre of genetic variability of the potato is Mexico, where important wild species that have been used extensively in modern breeding are found, such as the hexaploid* Solanum demissum, *as a source of resistance to the devastating late blight disease. Another relative native to this region,* Solanum bulbocastanum, *has been used to genetically engineer the potato to resist potato blight.*

Potatoes yield abundantly with little effort, and adapt readily to diverse climates as long as the climate is cool and moist enough for the plants to gather sufficient water from the soil to form the starchy tubers. Potatoes do not keep very well in storage and are vulnerable to molds that feed on the stored tubers, quickly turning them rotten. By contrast, grain can be stored for several years without much risk of rotting.

Growth and Cultivation

Potatoes are generally grown from seed potatoes – these are tubers specifically grown to be disease free and provide consistent and healthy plants. To be disease free, the areas where seed potatoes are grown are selected with care. In the USA this restricts production of seed potatoes to only 15 states out of the 50 states that grow potatoes. These locations are selected for their cold hard winters that kill pests and long sunshine hours in the summer for optimum growth. In the UK, most seed potatoes originate in Scotland in areas where westerly winds prevent aphid attack and thus prevent spread of potato virus pathogens. Potato growth has been divided into five phases.

During the first phase, sprouts emerge from the seed potatoes and root growth begins. During the second, photosynthesis begins as the plant develops leaves and branches. In the third phase stolons develop from lower leaf axils on the stem and grow downwards into the ground and on these stolons new tubers develop as swellings of the stolon. This phase is often (but not always) associated with flowering. Tuber formation halts when soil temperatures reach 27 °C (81 °F); hence potatoes are considered a cool-season crop. Tuber bulking occurs during the fourth phase, when the plant begins investing the majority of its resources in its newly formed tubers. At this stage, several factors are critical to yield: optimal soil moisture and temperature, soil nutrient availability and balance, and resistance to pest attacks. The final phase is maturation:

The plant canopy dies back, the tuber skins harden, and their sugars convert to starches.

New tubers may arise at the soil surface. Since exposure to light leads to greening of the skins and the development of solanine, growers are interested in covering such tubers. Commercial growers usually address this problem by piling additional soil around the base of the plant as it grows ("hilling", or in British English "earthing up"). An alternative method used by home gardeners and smaller-scale growers involves covering the growing area with organic mulches such as straw or with plastic sheets.

Correct potato husbandry can be an arduous task in some circumstances. Good ground preparation, harrowing, plowing, and rolling are always needed, along with a little grace from the weather and a good source of water. Three successive plowings, with associated harrowing and rolling, are desirable before planting. Eliminating all root-weeds is desirable in potato cultivation. In general, the potatoes themselves are grown from the eyes of another potato and not from seed. Home gardeners often plant a piece of potato with two or three eyes in a hill of mounded soil. Commercial growers plant potatoes as a row crop using seed tubers, young plants or microtubers and may mound the entire row. Seed potato crops are 'rogued' in some countries to eliminate diseased plants or those of a different variety from the seed crop.

Potatoes are sensitive to heavy frosts, which damage them in the ground. Even cold weather makes potatoes more susceptible to bruising and possibly later rotting, which can quickly ruin a large stored crop.

At harvest time, gardeners usually dig up potatoes with a long-handled, three-prong "grape" (or graip), i.e., a spading fork, or a potato hook, which is similar to the graip but with tines at a 90° angle to the handle. In larger plots, the plow is the fastest implement for unearthing potatoes. Commercial harvesting is typically done with large potato harvesters, which scoop up the plant and surrounding earth. This is transported up an apron chain consisting of steel links several feet wide, which separates some of the dirt. The chain deposits into an area where further separation occurs. Different designs use different systems at this point. The most complex designs use vine choppers and shakers, along with a blower system or "Flying Willard" to separate the potatoes from the plant. The result is then usually run past workers who continue to sort out plant material, stones, and rotten potatoes before the potatoes are continuously delivered to a wagon or truck. Further inspection and separation occurs when the potatoes are unloaded from the field vehicles and put into storage.

Immature potatoes may be sold as "new potatoes" and are particularly valued for taste. These are often harvested by the home gardener or farmer by "grabbling", i.e. pulling out the young tubers by hand while leaving the plant in place.

Potatoes are usually cured after harvest to improve skin-set. Skin-set is the process by which the skin of the potato becomes resistant to skinning damage. Potato tubers may be susceptible to skinning at harvest and suffer skinning damage during harvest and handling operations. Curing allows the skin to fully set and any wounds to heal. Wound-healing prevents infection and water-loss from the tubers during storage. Curing is normally done at relatively warm temperatures 50 to 60 °C (122 to 140 °F) with high humidity and good gas-exchange if at all possible.

Storage

Storage facilities need to be carefully designed to keep the potatoes alive and slow the natural process of decomposition, which involves the breakdown of starch. It is crucial that the storage area is dark, well ventilated and for long-term storage maintained at temperatures near 4 °C (39 °F). For short-term storage before cooking, temperatures of about 7 to 10 °C (45 to 50 °F) are preferred.

On the other hand, temperatures below 4 °C (39 °F) convert potatoes' starch into sugar, which alters their taste and cooking qualities and leads to higher acrylamide levels in the cooked product, especially in deep-fried dishes—the discovery of acrylamides in starchy foods in 2002 has led to many international health concerns as they are believed to be possible carcinogens and their occurrence in cooked foods is currently under study as a possible influence in potential health problems.

Under optimum conditions possible in commercial warehouses, potatoes can be stored for up to ten to twelve months. When stored in homes, the shelf life is usually only a few weeks.

If potatoes develop green areas or start to sprout, these areas should be trimmed before using. Trimming or peeling green areas are inadequate to remove copresent toxins, and such potatoes are no longer suitable as animal food.

Commercial storage of potatoes involves several phases: drying of surface moisture; a wound healing phase at 85% to 95% relative humidity and temperatures below 25 °C (77 °F); a staged cooling phase; a holding phase; and a reconditioning phase, during which the tubers are slowly warmed. Mechanical ventilation is used at various points during the process to prevent condensation and accumulation of carbon dioxide.

Yield

The world dedicated 18.6 million hectares in 2010 for potato cultivation. The average world farm yield for potato was 17.4 tonnes per hectare, in 2010. Potato farms in the United States were the most productive in 2010, with a nationwide average of 44.3 tonnes per hectare. United Kingdom was a close second.

New Zealand farmers have demonstrated some of the best commercial yields in the world, ranging between 60 to 80 tonnes per hectare, some reporting yields of 88 tonnes potatoes per hectare.

There is a big gap among various countries between high and low yields, even with the same variety of potato. Average potato yields in developed economies ranges between 38–44 tonnes per hectare. The two largest producers of potato, China and India which accounted for over a third of world's production in 2010, had yields of 14.7 and 19.9 tonnes per hectare respectively. The yield gap between farms in developing economies and developed economies represents an opportunity loss of over 400 million tonnes of potato, or an amount greater than 2010 world potato production. Potato crop yields are determined by factors such as the crop breed, seed age and quality, crop management practices and the plant environment. Improvements in one or more of these yield determinants, and a closure of the yield gap, can be a major boost to food supply and farmer incomes in the developing world.

Pests

Figure: *A potato ruined by late blight*

The historically significant *Phytophthora infestans* (late blight) remains an ongoing problem in Europe and the United States. Other potato diseases include *Rhizoctonia*, *Sclerotinia*, black leg, powdery mildew, powdery scab and leafroll virus.

Insects that commonly transmit potato diseases or damage the plants include the Colorado potato beetle, the potato tuber moth, the green peach aphid (*Myzus persicae*), the potato aphid, beetleafhoppers, thrips, and mites. The potato root nematode is a microscopic worm that thrives on the roots, thus causing the potato plants to wilt. Since its eggs can survive in the soil for several years, crop rotation is recommended.

Soil Insects

Soil insects, primarily wireworms and white grubs, can severely damage seed pieces and tubers. These insects feed on grass roots and should be considered as a serious threat when potatoes are to be grown in ground immediately following sod.

In these situations, a pre-plant broadcast application of a soil insecticide should be considered. For best results, treat after soil temperature at the six inch depth has reached 50øF. By this time soil insects should be active and nearer the surface.

The threat of damage by soil insects is lessened with time out of sod. Because some wireworms and grubs spend two or more years in the soil a problem may still occur. A planting-time soil treatment should be sufficient in these cases.

Above-Ground Pests

Colorado Potato Beetle

The common black and yellow-striped "potato bug", a very familiar insect, is the most serious pest of potatoes. Both the adult, or beetle, and the black-spotted, red larva feed on potato leaves. Their damage can greatly reduce yield and even kill plants.

In addition to potato, Colorado potato beetles can be a serious pest of tomato, eggplant, and pepper.

Colorado potato beetles overwinter in the soil as adults. The Colorado Potato Beetle is a yellow insect with alternating black and white strips down its back. They become active again in the spring and feed on weeds and volunteer on early planted potatoes. They will even enter the soil to attack emerging foliage. Female beetles lay batches of about two dozen orange-yellow eggs on the underside of the leaves. Each female can lay 500 or more eggs over a four to five week period. The eggs hatch in four to nine days and the larvae begin to feed on potato foliage. The larvae are humpbacked with two rows of black spots. The larvae usually feed in groups and damage can be severe. The larval stage lasts two to three weeks.

Full grown larvae move to the ground and change into an inactive or pupal stage. In five to 10 days the new adult beetles emerge. This insect can go from egg to adult in as little as 21 days. They feed for a few days, before egglaying begins. There are two full and occasionally a partial third generation each year. If foliar sprays are used, an effort should be made to treat after most eggs have hatched but before serious plant damage occurs.

This insect is notorious for development of resistance to insecticides over short periods of time. A rotation among different classes of insecticides is recommended to discourage resistance. There is a new commercially available strain of *Bacillius thuringiensis* (var *tenebrionis*) that are effective against small larvae (less than 1/4 inch) and should be applied at egg hatch or when larvae are first seen. A premature treatment may lose much of its effectiveness before the eggs hatch. An insect growth regulator, called Align, has also been released for control of this insect. It is an extract of the neem seed that prevent the insect from developing normally.

The Colorado potato beetle, *Leptinotarsa decemlineata*, can be easily confused with its close cousin the false potato beetle, *Leptinotarsa juncta*. While these two insects look nearly identical, only the Colorado potato beetle is a serious pest. While the adult false potato beetle has alternating black and white strips on its back as well, one of the white strips in the centre of each wing cover is missing and replaced by a light brown strip. The eggs are slightly larger and fewer found in a cluster. The humpbacked larva is similar, but with only one row of dark spots on each side. False potato beetles are frequently found feeding on some solanaceous weeds, such as horsenettle, but do no growth and reproduction occurs when feeding on potato.

Aphids

Winged aphids may move into potato fields in significant numbers. These migrants settle on the leaves and begin to remove plant sap. During this period they are also capable of producing large numbers of wingless aphids that in turn place an even greater stress on plants.

Distorted leaves and "sticky" leaf surfaces are signs of aphid infestation. Natural enemies and diseases can often keep aphid populations under control. Limiting the use of insecticides other than *Bacillius thuringiensis* products will conserve predators and parasites that help keep aphid populations under control.

Potato Leafhopper

Potato leafhoppers are wedge-shaped, 1/8-inch long, green, active insects. They use their piercing-sucking mouthparts to remove sap from the potato leaf. These small insects fly readily.

Because of their small size and habit of feeding on the underside of the leaf, they are easily overlooked.

The symptom of leafhopper activity is more apparent - a triangular brown spot at the tip of the leaf. Similar triangles may appear at the end of each lateral veinlet or the entire margin may roll upward as though scorched. These symptoms are known as "hopperburn". Other conditions may produce similar symptoms. Check the underside of leaves for the tiny leafhoppers to confirm that they are the cause of the problem.

Potato leafhoppers do not overwinter in Kentucky. Winds carry them into the state each year from the Gulf Coast. They generally appear between May 25 and June 5. Within a few days after mating, females lay their eggs in the stems and larger leaf veins of succulent plants. The eggs hatch in about 10 days into the immature or nymphal stage. The light green, wedge-shaped nymphs are smaller than the adults and do not have wings. Both stages are very active. Adults jump or fly readily when disturbed while the nymphs run sideways across the leaf and over the edge.

Development from egg to adult takes about three weeks in warm weather. Very large leafhopper populations can build up in a short time during the summer.

Occasional Pests

A variety of other insects may be numerous at times. Control decisions should be based on the extent of the infestation in the field and the severity of damage.

Blister beetles (shown right) are narrow, elongate insects. They may be found feeding in clusters on potato leaves. Infestations are usually localized within fields and treatment of an entire field is seldom necessary. Rarely do blister beetles do enough damage to cause any yield loss.

Hornworms are easily identified and will feed on potato foliage. These worms can consume large amounts of foliage but severe infestations are not likely to occur. Hormworms can easily be controlled using any of the insecticides containing Bacillus thuringiensis *var kurstaki. Rarely do hormworms cause losses to potato yields.*

Flea beetles cause the small shot-hole damage to leaves. These tiny beetles overwinter as adults and may appear in fields very early in the season and cause serious damage to young plants. Sevin provides very good control of flea beetles when they are numerous on small plants.

Systemic Insecticides

Planting time use of systemic insecticides allows protection against moderate levels of soil insects plus activity against leaf-feeding pests such as aphids, leafhoppers and flea beetles. Insecticidal effects are generally greatest during the early part of the growing season.

These insecticides are highly toxic and should be used with extreme care. Read the label thoroughly and use protective equipment. These materials require specialised application equipment.

Foliar Sprays

A variety of insecticides may be used on potatoes to control insect pests. Repeated sprays may be needed in some years, especially against late season pests.

Colorado Potato Beetle Resistance Management

The Colorado potato beetle is notorious for its ability to rapidly develop resistance to insecticides that are used repeatedly to control it. This has been a serious problem on the east coast for some time, and is becoming more of a problem in Kentucky. With a more limited number of available insecticides, some homeowners feel they have exhausted their control options.

Resistance develops more rapidly to an insecticide when that insecticide is used repeatedly as the only control measure. Often over use of one insecticide may favour the development of resistance to other insecticides in the same chemical class. Insecticides in the same chemical class usually have the same *mode of action*, i.e. the same method of killing the insect. Consequently, to delay or prevent resistance it is important to avoid repeated usage of one particular insecticide by rotating the insecticides used. Rotation needs to be among different classes of insecticides.

Other control measures such as hand picking of adult beetles and immature stages will also aid to delay the development of resistance. Resistance by Colorado potato beetles should be managed on a field to field basis. While they may be resistance to one insecticide in a particular field, those in other fields within the same county may not have developed resistance to that insecticide.

Pesticides

During the crop year 2008, many of the certified organic potatoes produced in the United Kingdom and certified by the Soil Association as organic were sprayed with a copper pesticide to control potato blight (*Phytophthora infestans*). According to the Soil Association, the total copper that can be applied to organic land is 6 kg/ha/year.

According to an Environmental Working Group analysis of USDA and FDA pesticide residue tests performed from 2000 through 2008, 84% of the 2,216 tested potato samples contained detectable traces of at least one pesticide. A total of 36 unique pesticides were detected on potatoes over the 2,216 samples, though no individual sample contained more than 6 unique pesticide traces, and the average was 1.29 detectable unique pesticide traces per sample. The average quantity of all pesticide traces found in the 2,216 samples was 1.602 ppm. While this is a very low value of pesticide residue, it is the highest amongst the 50 vegetables analyzed.

Sweet Potato

The sweet potato (*Ipomoea batatas*) is a dicotyledonous plant that belongs to the family Convolvulaceae. Its large, starchy, sweet-tasting, tuberous roots are a root vegetable. The young leaves and shoots are sometimes eaten as greens. Ipomoea batatas is native to the tropical regions in America. Of the approximately 50 genera and more than 1,000 species of Convolvulaceae, *I. batatas* is the only crop plant of major importance—some others are used locally, but many are poisonous. The sweet potato is only distantly related to the potato (*Solanum tuberosum*) and does not belong to the nightshade family.

The genus *Ipomoea* that contains the sweet potato also includes several garden flowers called morning glories, though that term is not usually extended to *Ipomoea batatas*. Some cultivars of *Ipomoea batatas* are grown as ornamental plants; the name "tuberous morning glory" may be used in a horticultural context. The plant is a herbaceous perennial vine, bearing alternate heart-shaped or palmately lobed leaves and medium-sized sympetalous flowers. The edible tuberous root is long and tapered, with a smooth skin whose colour ranges between yellow, orange, red, brown, purple, and beige. Its flesh ranges from beige through white, red, pink, violet, yellow, orange, and purple. Sweet potato varieties with white or pale yellow flesh are less sweet and moist than those with red, pink or orange flesh.

In certain parts of the world, sweet potatoes are locally known by other names. In New Zealand English, the Mâori term *kûmara* (also spelled *kûmera*) is commonly used. Although the soft, orange sweet potato is often called a *"yam"* in parts of North America, the sweet potato is botanically very distinct from a genuine yam (*Dioscorea*), which is native to Africa and Asia and belongs to the monocot family Dioscoreaceae. To add to the confusion, a different crop plant, the oca, *Oxalis tuberosa* (a species of woodbind), is called a "yam" in many parts of Polynesia, including New Zealand. To prevent confusion, the United States Department of Agriculture requires sweet potatoes labelled as "yams" to also be labelled as "sweet potatoes". The sweet potato is North Carolina's state vegetable.

Origin, Distribution and Diversity

Figure: *Sweet potatoes in the field*

The origin and domestication of sweet potato is thought to be in either Central America or South America. In Central America, sweet potatoes were domesticated at least 5,000 years ago. In South America, Peruvian sweet potato remnants dating as far back as 8000 BC have been found.

Daniel F. Austin postulated that the centre of origin of *I. batatas* was between the Yucatán Peninsula of Mexico and the mouth of the

Orinoco River in Venezuela. The 'cultigen' had most likely been spread by local people to the Caribbean and South America by 2500 BC. Strong supporting evidence was provided that the geographical zone postulated by Austin is the primary centre of diversity. The much lower molecular diversity found in Peru–Ecuador suggests this region should be considered as secondary centre of sweet potato diversity.

The sweet potato was also grown before western exploration in Polynesia. Sweet potato has been radiocarbon-dated in the Cook Islands to 1000 AD, and current thinking is that it was brought to central Polynesia around 700 AD, possibly by Polynesians who had travelled to South America and back, and spread across Polynesia to Hawaii and New Zealand from there. It is possible, however, that South Americans brought it to the Pacific, although this is unlikely as it was the Polynesians who had a strong maritime tradition and not the native South Americans. The theory that the plant could spread by floating seeds across the ocean is not supported by evidence. Another point is that the sweet potato in Polynesia is the cultivated *Ipomoea batatas*, which is generally spread by vine cuttings and not by seeds.

Sweet potatoes are now cultivated throughout tropical and warm temperate regions wherever there is sufficient water to support their growth. For example, sweet potatoes were introduced as a food crop in Japan, and by 1735 was planted in Shogun Tokugawa Yoshimune's private garden. It was also introduced to Korea in 1764.

According to the Food and Agriculture Organisation (FAO) statistics, world production in 2004 was 127 million tonnes. The majority comes from China, with a production of 105 million tonnes from 49,000 km^2. About half of the Chinese crop is used for livestock feed.

Per capita production is greatest in countries where sweet potatoes are a staple of human consumption, led by Papua New Guinea at about 500 kg per person per year, the Solomon Islands at 160 kg, Burundi and Rwanda at 130 kg and Uganda at 100 kg.

About 20,000 tonnes of sweet potatoes are produced annually in New Zealand, where sweet potato is known by its Mâori name, *kûmara*. It was a staple food for Mâori before European contact.

In the U.S., North Carolina, the leading state in sweet potato production, provided 38.5% of the 2007 U.S. production of sweet potatoes. In 2007, California produced 23%, Louisiana 15.9%, and Mississippi 19% of the U.S. total.

The town of Opelousas, Louisiana's "Yambilee" has been celebrated every October since 1946. The Frenchmen who established the first settlement at Opelousas in 1760 discovered the native Atakapa, Alabama, Choctaw, and Appalousa tribes eating sweet potatoes. The sweet potato became a favourite food item of the French and Spanish settlers and thus continued a long history of cultivation in Louisiana.

Mississippi has about 150 farmers growing sweet potatoes on about 8,200 acres (30 km), contributing $19 million to the state's economy. Mississippi's top five sweet potato producing counties are Calhoun, Chickasaw, Pontotoc, Yalobusha, and Panola. The National Sweet Potato Festival is held annually the entire first week in November in Vardaman (Calhoun County), which proclaims itself as "The Sweet Potato Capital".

The town of Benton, Kentucky, celebrates the sweet potato annually with its Tater Day Festival on the first Monday of April. The town of Gleason, Tennessee, celebrates the sweet potato on Labour Day weekend with its Tater Town Special.

Sweet potatoes became popular very early in the islands of the Pacific Ocean, spreading from Polynesia to Japan and the Philippines. One reason is that they were a reliable crop in cases of crop failure of other staple foods because of typhoon flooding. They are featured in many favourite dishes in Japan, Taiwan, the Philippines, and other island nations. Indonesia, Vietnam, India, and some other Asian countries are also large sweet potato growers. Sweet potato, also known as *kelang* in Tulu is part of Udupi cuisine. Uganda (the second largest grower after China), Rwanda, and some other African countries also grow a large crop which is an important part of their peoples' diets. The New World, the original home of the sweet potato, grows less than three percent (3%) of the world's supply. Europe has only a very small sweet potato production, mainly in Portugal. In the Caribbean, a variety of the sweet potato called the *boniato* is popular. The flesh of the *boniato* is cream-coloured, unlike the more popular orange hue seen in other varieties. *Boniatos* are not as sweet and moist as other sweet potatoes, but many people prefer their fluffier consistency and more delicate flavour.

Figure: *Sweet Potato Harvest.*

Sweet potatoes have been an important part of the diet in the United States for most of its history, especially in the Southeast. From the middle of the 20th century, however, they have become less popular. The average per capita consumption of sweet potatoes in the United States is only about 1.5–2 kg (3.3–4.4 lb) per year, down from 13 kg (29 lb) in 1920. Southerner Kent Wrench writes: "The Sweet Potato became associated with hard times in the minds of our ancestors and when they became affluent enough to change their menu, the potato was served less often."

Cultivation

Producers (in Million Tonnes) Data for Year 2011	
China	81.7
Uganda	2.8
Nigeria	2.8
Indonesia	2.0
Tanzania	1.4
Vietnam	1.3
India	1.1
United States	1.0
World	106.5

The plant does not tolerate frost. It grows best at an average temperature of 24 °C (75 °F), abundant sunshine and warm nights. Annual rainfalls of 750–1,000 mm (30–39 in) are considered most suitable, with a minimum of 500 mm (20 in) in the growing season. The crop is sensitive to drought at the tuber initiation stage 50–60 days after planting, and it is not tolerant to water-logging, as it may cause tuber rots and reduce growth of storage roots if aeration is poor.

Depending on the cultivar and conditions, tuberous roots mature in two to nine months. With care, early-maturing cultivars can be grown as an annual summer crop in temperate areas, such as the northern United States. Sweet potatoes rarely flower when the daylight is longer than 11 hours, as is normal outside of the tropics. They are mostly propagated by stem or root cuttings or by adventitious roots called "slips" that grow out from the tuberous roots during storage. True seeds are used for breeding only.

They grow well in many farming conditions and have few natural enemies; pesticides are rarely needed. Sweet potatoes are grown on a variety of soils, but well-drained, light- and medium-textured soils with a pH range of 4.5-7.0 are more favourable for the plant. They can be grown in poor soils with little fertilizer. However, sweet potatoes are very sensitive to aluminum toxicity and will die about six weeks after planting if lime is not applied at planting in this type of soil. Because they are sown by vine cuttings rather than seeds, sweet potatoes are relatively easy to plant. Because the rapidly growing vines shade out weeds, little weeding is needed. In the tropics, the crop can be maintained in the ground and harvested as needed for market or home consumption. In temperate regions, sweet potatoes are most often grown on larger farms and are harvested before first frosts.

Figure: *The softer, orange-fleshed variety of sweet potato*

Figure: *Naturalized sweet potato in the Philippines (locally known as* camote*)*

In the Southeastern United States, sweet potatoes are traditionally cured to improve storage, flavour, and nutrition, and to allow wounds on the periderm of the harvested root to heal. Proper curing requires drying the freshly dug roots on the ground for two to three hours, then storage at 85–90 °F (29–32 °C) with 90 to 95% relative humidity from five to fourteen days. Cured sweet potatoes can keep for thirteen months when stored at 55–59 °F (13–15 °C) with >90% relative humidity. Colder temperatures injure the roots.

Yields of Sweet Potato Crop

In 2010, the world average annual yield for sweet potato crop was 13.2 tonnes per hectare. The most productive farms of sweet potato breeds were in Senegal, where the nationwide average annual yield was 33.3 tonnes per hectare. Yields as high as 80 metric tonnes per hectare have been reported from farms of Israel.

Sweet Potato Storage

The sweet potato (*Ipomoea batatas*) is a very important crop for subsistence farmers in Africa and other developing countries. Due to its relatively short growing period, tolerance to drought and high yield from poor soils, it is used as a famine reserve for many of these households. However it is a highly perishable food source that is susceptible to destruction by microorganisms, metabolic spoilage, physical destruction and pests. As such they are not generally stored for extended periods of time following harvest. This is a major barrier for the optimal use of the crop and causes a lot of waste.

The majority of subsistence farmers who eat sweet potato do not utilise a storage technique of any kind, instead using a method of continuous cultivation and leaving the tubers in the ground until they are ready to be used. While less common, simple methods of storage do exist. The common ones include pit, clamp, and indoor storage and all extend the storage time of sweet potatoes by some degree. Maintaining proper storage temperature is probably the most significant barrier to good storage in subsistence farmers. Pretreatment of sweet potato can help to minimize risk of losses. Curing can toughen the skin and heal minor physical damages while drying can reduce spoilage and inactivate metabolic degradation.

Optimum storage of sweet potatoes occurs at 12–16 °C (54–61 °F), 85–90% relative humidity (RH) and requires proper ventilation to remove excess carbon dioxide (CO2) and bring in oxygen (O2) for respiration. At these conditions sweet potatoes have been shown to last 5 months to a maximum of a year compared to 2–3 months normally.

Susceptibility to Spoilage

Sweet potatoes are a highly perishable crop and difficult to store for extended periods of time. This is mainly due to their high moisture content, metabolic activity following harvesting and thin, permeable skin. Following harvest, sweet potatoes are susceptible to spoilage by physical, physiological, pathological, pests and environmental means.

Physical

Physical losses refer to sweet potatoes damaged by mechanical force during harvest, handling, and transportation of the tubers. This includes injury by way of cuts, bruising, and skinning. Delicate skin and poor harvesting and handling techniques are the main cause of these issues. Damage to tubers by physical means can be minimized careful handling and storing in boxes and cartons instead of cloth sacks.

Physiological

Physiological processes within the plant can contribute highly to spoilage. Natural transpiration causes water loss and results in an unappealing pithy texture. Respiration also continues to occur after harvesting and a buildup of CO2 around the crop can cause early spoilage. Sprouting will also occur during long term storage, which is useful for planting but will utilise nutrients in the tuber. Sprouting can be inhibited by preharvest spraying with maleic hydrazide, treatment of tubers with methyl ester of napthalene acetic acid (MENA) in acetone on paper spread between tubers and 0.5–4.0% thiourea solution.

Environmental

The length of time that sweet potatoes may be stored is highly dependent on temperature and relative humidity. Storage in temperatures less than 10 °C (50 °F) will cause chill injury in sweet potatoes. This is characterized by discolouration of the insides of the tuber, which can range from brown to black instead of the usual orange. Formation of hard areas also occurs that remain even after cooking.

Other signs of chill injury include production of off flavours, internal cell structure breakdown and increased rotting due to reduced skin formation, which allows fungal infection. On the other end of the scale, temperatures above 16 °C (61 °F) will cause a rise in the rates of respiration and sprouting in tubers as well as increased activity by microorganisms.

Relative humidity under 80% will cause excessive water loss while RH above 90% will cause condensation to form, which accelerates rotting.

Ideal Storage Conditions

To minimize potential spoilage, precise storage conditions should be met. Sweet potatoes should be stored at between 12 and 16 °C (54 and 61 °F) with a RH of 85-90%. Proper ventilation of the storage area is a key factor in maximizing potential storage time. Respiration causes the production of CO2, which will act to increase the rate of spoilage. Sweet potatoes can convert approximately 57dm3 of oxygen/ton/day to carbon dioxide and need ventilation to compensate for this. The O2 in the storage environment should not fall below 7% and CO2 must not exceed 10%. Due to lack of proper equipment, maintaining adequate temperature conditions are the largest hurdle that farmers in a developing nation will have to face.

Not all sweet potatoes are suitable for storage. Tubers that have been mechanically damaged, infected with pests or pathogens, or are rotten or sprouting should not be kept. These factors will increase rate of spoilage and can increase losses in the rest of the stored crop. Under these optimum conditions sweet potatoes have been shown to keep for 5 months up to a max of about a year.

Storage Methods

In developed countries proper refrigeration technology is used to keep sweet potatoes for long periods of time. Many poor farmers do not have access to refrigeration or even electricity so must rely on other storage methods. There are several methods of storage that only require simple and cheap building mCaterials. The most commonly used are pit storage, clamp storage and indoor storage. These techniques are used only sporadically however as most farmers practice sequential harvesting in which tubers are left in-ground until needed.

In-Ground

In subsistence farming, sweet potatoes are commonly left in the ground and eaten or sold directly following harvest, this is called piece-meal or sequential harvesting. Sweet potatoes are delicate and easily damaged. In-ground storage is used to protect the tubers while reducing the work required to set up storage facilities. In areas with reliable rainfall patterns farmers are able to maintain a supply of fresh sweet potatoes for most of the year by continuous cultivation and harvesting. In drier places however, such as northern Uganda, this is not a practical solution. This is not a long-term option and the plants will degrade after a short period of time. There are numerous disadvantages to this method as it takes up field space that could otherwise be used for growing more crops.

Pit Storage

This is a simple, cheap method that involves digging a hole in the ground for storing the potatoes. Pit storage differs from in-ground as tubers are collected and kept together and considerations are made to control the storage environment.

Construction of the pit can vary according to what materials are available but commonly used are; grass, soil, wood, lime, sawdust and ash. Grass is typically used to line the bottom and sides of the pit in order to insulate against temperature change and absorb moisture. Before usage, grass should be flamed in order to destroy any pests that may be hiding within. Soil is used to seal the roof of the pit and as filler. Wood and plant material can be used to strengthen walls as well as create a roof covering the pit. Roofs built over the pit structure are beneficial in that they can help keep rain out and provide shade to lower temperature. Lime may have some effectiveness in absorbing CO2 and removing it form the environment. Sawdust is used as a cushioning material and to help control condensation on the tubers. Wood ash can applied to potatoes prior to storage and has shown some effectiveness in protecting against insect attacks and mould.

In order to prevent rotting the pit should ideally be dug into a slope to promote drainage. It should also be situated at least 0.35 metres (1.1 ft) above the water table during dry season and must remain above during wet/rainy season and be approximately 0.5 metres (1.6 ft) below the surface.

There can be ventilation problems with this storage method. Commonly the pit will be sealed shut in order to maintain a good RH. The draw back to this is that CO2 will collect and cause spoilage. To help this, a sizable headspace should be maintained in the pit to help promote airflow.

Pits can be reused but they should be cleaned, the soil turned over and disinfected with fire or sulphur to rid the area of any microorganisms.

Losses are still fairly high for this technique, though it has been shown in some instances to keep roots for 2–4 months.

Clamp and Mound Storage

Another simple and low cost method, clamp storage consists of covered piles of sweet potatoes. After selecting for the tubers in the best condition, they are stacked in a heap on a layer of grass and covered in layers of grass and soil. As with pit storage ash, lime and sawdust can be used for added effect. The piles may be made at ground level or in shallow or deep trenches. Drainage should be considered and ruts may be made in the ground to lead off water. The clamps may be covered by a roof or kept in a building for added protection. To minimize losses due to respiration a ventilation shaft can be added. Results are fairly poor with this technique and estimated storage time is 2–3 months.

Indoor Storage

Sweet potatoes may be harvested and stored in building. This could include in the living area or in a granary built specifically to store produce. In home storage is typically done in straw woven baskets, cloth bags or wooden boxes. Baskets and boxes have been shown to be more effective at minimizing mechanical damage. If possible tubers should be kept off of the ground to keep them away from rodents and other pests. This is an effective technique for maintaining proper ventilation though depending on the type of building maintaining proper storage temperature and RH may be difficult.

Granaries or other storage buildings typically consist of a round hut with walls made of straw, mud, clay and wood and a conical straw roof. These are commonly supported above ground by a system of legs to keep the crop dry and away from animals, rodents and pests.

Pretreatment of Sweet Potatoes

Typically sweet potatoes are stored and eaten fresh. However there are some simple methods used to increase their storability that can be used in conjunction with other storage methods. Drying and curing of sweet potatoes are two common methods to prepare them prior to storage.

Drying of tubers is done on those that are too damaged to be stored fresh but still have edible material on them. This involves slicing tubers to a thickness of approximately 2–4 mm and then laying them out in the sun for four days or until they are rid of most of their moisture. During drying the potatoes can be covered in prickly bushes or thorns to ward off animals. Dried slices can be kept in-doors or in raised silos until eaten. Drying removes moisture, reduces bacterial growth, and inactivates metabolic processes and enzymatic decomposition.

Curing is a technique done to toughen up the outer layer of skin (periderm). It is a measure that prevents against excessive moisture loss, entry of microorganisms into the plant and even facilitates healing of mechanically damaged tubers. Curing is done by exposing the whole potatoes to a moderately high temperature for several days immediately following harvest.; 30–32 °C (86–90 °F) and 80–95% RH for 4–10 days has been found to be adequate curing conditions A common method to cure without use of incubation equipment is to wrap tubers in black polythene sheets and leave out in the sun for 5 days. The black sheet keeps moisture in and collects heat to reach necessary conditions for curing. Immediately following curing the temperature must be dropped quickly in order to avoid sprouting damage.

Pathological and Pests

Sweet potatoes are susceptible to a number of fungi, bacteria and viruses. Infection can cause the production of toxins, pathogenic bacteria, bitter flavours, tissue breakdown, and surface blemishes. Two common sources of infection are from *Rhizopus oryzae* (soft rot) and *Botryodiplodia* (Java black rot). These are responsible for 78% of sweet potato infections in Bangladesh. The pest of greatest concern for farmers is the sweet potato weevil (*Cylas spp.*). This can enter the tuber prior to harvesting and proliferate in storage and has been known to cause up to 50% loss in yields. Simple techniques can be used to minimize the threat of the sweet potato weevil. It has been shown that storage of the potatoes in temperatures under 20 °C (68 °F) can kill up to 89.5% of weevils in an infected crop. Dipping of tubers in hot water also been shown as an effective intervention to control infestations . Damage to tube by physical and physiological means can increase chance of pathological infection and pests.

It is useful to select cultivars that have resistances to the more common insects and diseases and employ crop rotation. Planting sweet potatoes in the same field once every 3 years helps to reduce pests and diseases. Crops most often rotated with sweet potatoes include tobacco, soybeans, corn, cotton wheat, cucumbers, peanuts, and rye. Insect monitoring is important for managing both foliar-feeding and root-feeding insects. Routine field scouting either by the grower, a family member, or a professional scout and the use of light traps and seed baits in the soil will identify which pests are present and what course of action is necessary. Should chemical insect control be warranted, the grower should follow all applicable directions, restrictions, and precautions of the EPA registered label.

In addition to adhering to the steps outlined above, it is beneficial to record every pest problem observed and diagnosed in each field or storage lot and keep those records available for future reference. Assistance in diagnosing pest problems is available from county agricultural extension agents. These agents are often able to identify the cause of problems quickly. In addition, many crop consultants and private laboratories can help with these and related services.

Insect Pests

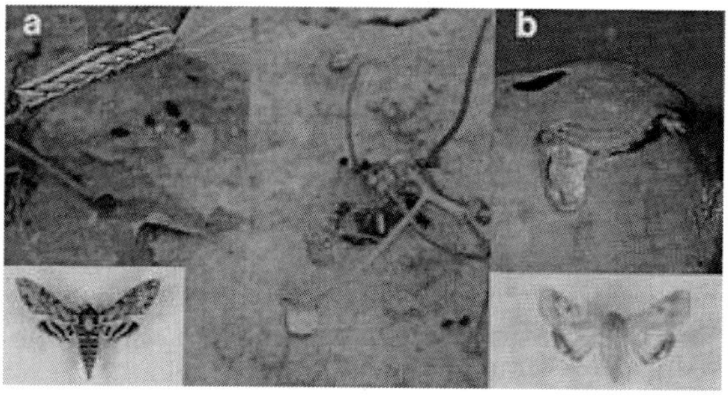

Figure: *(a) Sweet potato hornworm, Agrius cingulatus (Fab.); (b) corn earworm, Heliothis zea (Boddie)*

Lepidoptera larvae feed on foliage leaving small to large, irregular holes. In plant beds and newly set fields damage may be serious. After harvest, larvae may continue feeding on sweet potatoes left in the field and in storage. Management: Apply insecticide to plant beds. Cuttings should be free of insects before planting. Remove harvested sweet potatoes from the field immediately.

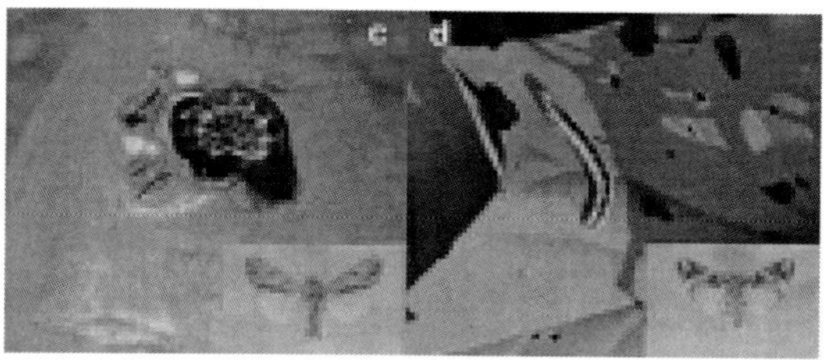

Figure: *(c) Southern armyworm, Spodoptera eridania (Cramer); (d) yellowstriped armyworm, Spodoptera omithogalli (Guenee)*

Two Spotted Spider Mite, *Tetranychus urticae* Koch. Immature mites and adults feed on underside of leaves and may cause stippling on leaves which appear yellowish, bronzed or burned. Mites favour hot, dry conditions, and often migrate from adjacent vegetation. Management: Minimize application of approved insecticides. Chemical control is usually not necessary.

Figure: *(a) mottled tortoise beetle, Deloyola guttata (Olivier); (b) striped tortoise beetle, Agrioconata bivittata (Say)*

Tortoise Beetle adults and larvae that feed on sweet potato foliage include: mottled tortoise beetle, striped tortoise beetle and (not shown) argus tortoise beetle, *Chelymorpha cassidea* (Fab.); black-legged tortoise beetle, *Jonthonata nigripes* (Oliver); and golden tortoise beetle, *Metriona bicolour* (Fab.). Leaves of infested plants are riddled with large round holes. Generally damage by tortoise beetles only threatens seedlings or newly set plants. Management: Isolate plant beds and control morning glory. Apply approved insecticides to young plants if needed. Control beetles on plants around plant beds and fields.

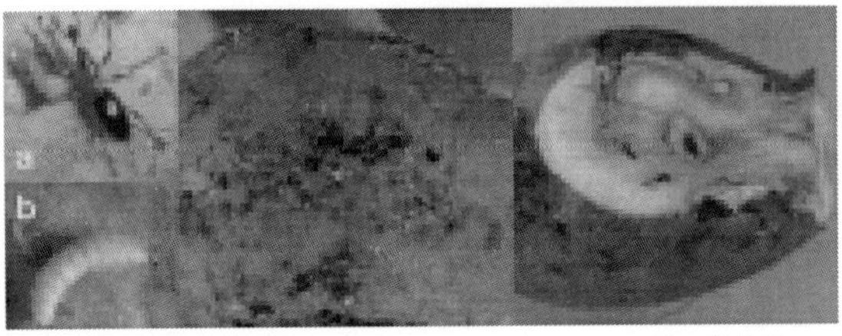

Figure: *(a) Adult sweet potato weevil, (b) larvae*

Sweet Potato Weevil, *Cylas formicarius elegantulus* (Summers) is the most serious, worldwide pest. Its northern distribution is on the Outer Banks near Wilmington, NC where it feeds on seaside morning glory. Adults and larvae feed on foliage but prefer stems and roots. Infested sweet potatoes are riddled with small holes and galleries especially in the stem end. They turn bitter and are unfit for consumption by either humans or livestock. Management: Use only seed and plant stocks produced in approved weevil-free areas. All purchased stocks, including out-of-state, must be certified. Use pheromone traps in plant beds, storage houses, and in fields to detect sweet potato weevil. Some varietal tolerance exists. Chemical control is difficult at best.

Sweet Potato Flea Beetle, Chaetocnema confinis Crotch. Adult beetles spend the winter in debris, along fencerows, and at the edges of wooded areas. In the spring, eggs are laid in the soil near host plants. There are several generations per year. Adults feed on foliage leaving channels on the upper leaf surfaces. Larvae feed on roots etching shallow, winding, sunken trails on the surface, which enlarge, darken and split. Management: Control weeds along field margins and plow under crop debris. Use resistant or tolerant varieties. In fields with a history of infestation use a pre-plant soil insecticide.

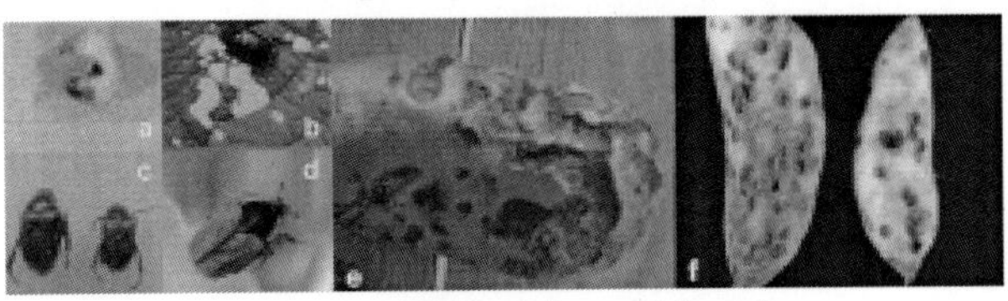

Figure: *Whitefringed beetle*

Whitefringed Beetle, *Graphognathus* spp..Larvae feed on roots causing damage similar to that of wireworms and white grubs. Only flightless, female adults occur and feed at the base of plants leaving scars on the stem. They are most active in July and August and produce eggs without mating. Management: Avoid infested fields and rotate crops. Limited control may be achieved with resistant varieties and use of soil insecticides.

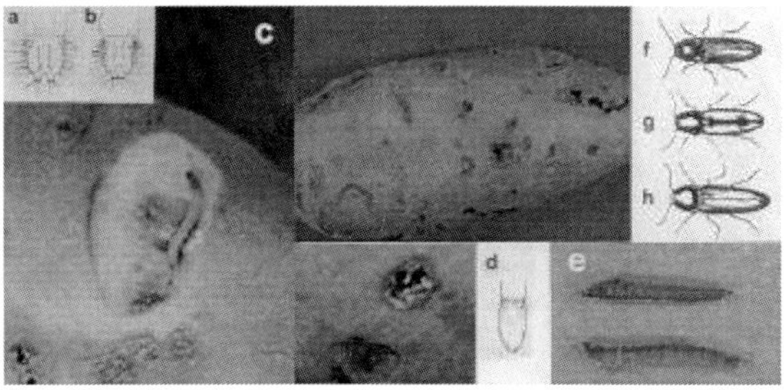

Figure: *(a,c,g)Tobacco wireworm, Conoderus vespertinus (Fab.); (b,f) southern potato wireworm, Conoderus falli Lane and (d,e,h) corn wireworm, Melanotus communis (Gyllenhal)*

Wireworms leave small, irregular, shallow or deep holes in the surface of sweet potato roots. Larvae are identified by differences in the last abdominal segment. Wireworm adults lay their eggs in grassy, undisturbed soil. Wireworms may be detected prior to planting using corn, wheat, or oatmeal bait stations. Management: Avoid land previously not in row crops, in sod, or fallow. If necessary, broadcast and incorporate a pre-plant insecticide or use a granular material at root swell.

Diabrotica and Systena spp. (not illustrated). Cucumber beetles and flea beetles may damage sweet potatoes. Adults produce irregular holes in foliage and larvae eat small holes through the periderm and form irregular cavities under the skin. Management: Treatment for wireworms may provide some control. Foliar sprays control adult beetles.

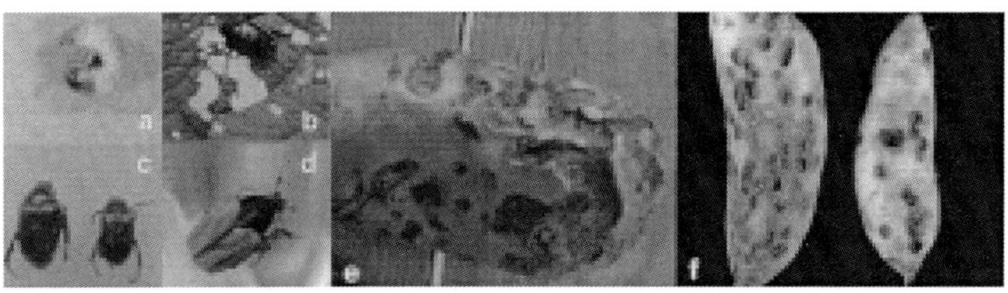

Figure: *(a) large, shallow, irregular damage on the surface (b,e) Japanese beetle, Popoillia japonica Newman; (c,f) spring rose beetle, Strigoderma arboricola Fab.; and (d) green June beetle, Cotinis nitida (l.)*

White grubs cause large, shallow, irregular damage on the surface of sweet potatoes. Species in North Carolina include Japanese beetle, *Popoillia japonica* Newman; spring rose beetle, *Strigoderma arboricola* Fab.; and green June beetle, *Cotinis nitida* (l.). Adults lay eggs in grassy areas. Management.

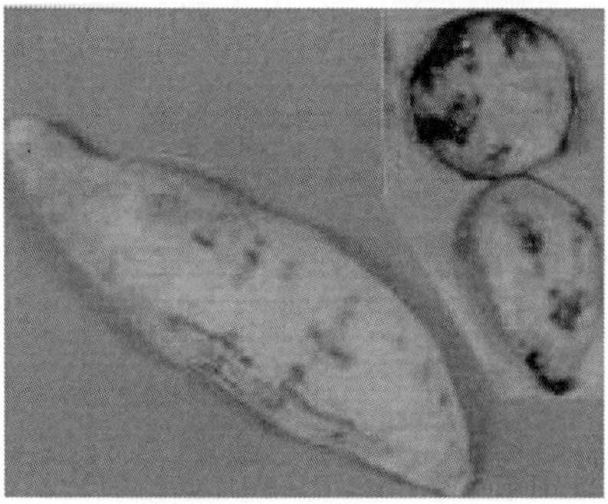

Figure: *Nematode*

Nematode problems may be caused by the southern root-knot nematode, *Meloidogyne incognita*; the northern root-knot nematode, *Meloidogyne arenaria*; the reniform nematode, *Rotylenchus reniformis*; and others. Symptoms include plant stunting, yellowing, leaf fall, reduced yields, and roots with cracks and black, partially decayed surfaces. Symptoms are much worse under drought conditions. Management: Assay fields for nematodes in the fall. Avoid problem fields. Treat infested soils in the row with a nematicide prior to or during planting. Use resistant cultivars.

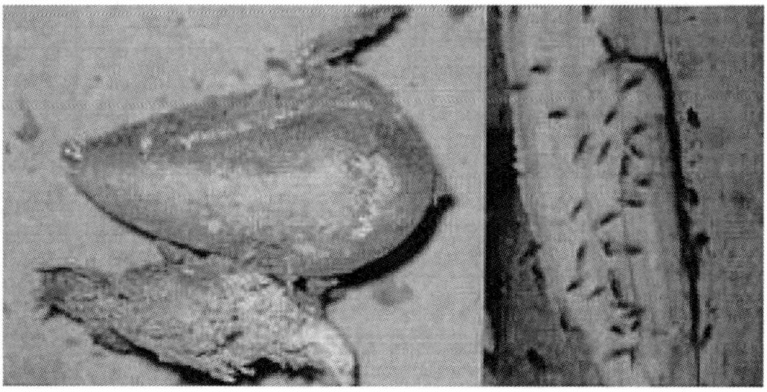

Figure: *Fruitfly*

Fruit Fly, *Drosophila* spp. Fruit flies may be a nuisance in storage houses when sweet potatoes decay due to other causes such as souring, chilling, and Rhizopus soft rot. Fruit flies feed on decaying vegetables. Maggots may be seen in decaying roots. Fruit flies may become established in cull piles and spread to the storage house. They do not cause rots. Management: Harvest, cure and store only sound sweet potatoes. Dispose of culls, inspect the storage house and use traps. If necessary spray with an appropriate insecticide.

The following Insect Management Program for sweet potatoes was developed by Dr. Kenneth Sorensen, Entomologist, NC State University and Bill Jester, Area Specialised Agent Commercial Vegetables and Fruit, NC Cooperative Extension Service.

Total Sweet Potato Insect Management Scheme

- Develop field histories based on soil baiting (February) and previous experience. Avoid fields with white fringed beetle.
- Use Admire in plant beds.
- Use Lorsban preplant. Fumigate with Telone.
- Scout fields biweekly after transplanting.
- Apply Admire as a soil drench 30-45 days after transplanting.
- Increase monitoring in early June. Eg: light traps, sweep nets.
- Begin foliar insecticide sprays using suggested action thresholds or follow the Bifenthrin (Brigade) IPM Program – view Bifenthrin IPM Program Chart
- Continue sprays every 7 to 14 days until September.

Suggested Action Threshold

- White Fringed Beetle – soil baits – any beetles, do not plant
- Grubs – 3 beetles in white water pails or 5 beetles in Japanese beetle traps
- Wireworm - larvae in corn baits, 5 beetles in yellow sticky traps or black light trap.
- Sweet Potato Flea Beetle – 5 beetles in yellow sticky traps or net sweeps, 5% damage early
- Cucumber Beetles – 10 beetles in traps, 10% damage

Foliar Insecticides

- Sevin
- Thiodan – Phaser
- Imidan
- Intrepid
- Spintor
- Confirm
- BT
- Rimon

Suggested Insecticide Program

- Lorsban Preplant
- Fumigants
- Admire in Plant Bed
- Admire Drench 30 to 45 Days After Planting
- Foliar Sprays Every 7 to 14 Days

6

Leafy Vegetables and Salad Crops

Leaf vegetables, also called potherbs, greens, vegetable greens, leafy greens or salad greens, are plant leaves eaten as a vegetable, sometimes accompanied by tender petioles and shoots. Although they come from a very wide variety of plants, most share a great deal with other leaf vegetables in nutrition and cooking methods. Nearly one thousand species of plants with edible leaves are known. Leaf vegetables most often come from short-lived herbaceous plants such as lettuce and spinach. Woody plants whose leaves can be eaten as leaf vegetables include *Adansonia*, *Aralia*, *Moringa*, *Morus*, and *Toona* species.

The leaves of many fodder crops are also edible by humans, but usually only eaten under famine conditions. Examples include alfalfa, clover, and most grasses, including wheat and barley. These plants are often much more prolific than more traditional leaf vegetables, but exploitation of their rich nutrition is difficult, primarily because of their high fibre content. This obstacle can be overcome by further processing such as drying and grinding into powder or pulping and pressing for juice.

Leaf vegetables contain many typical plant nutrients, but since they are photosynthetic tissues, their vitamin K levels in relation to those of other fruits and vegetables, as well as other types of foods, is particularly notable. The reason is that phylloquinone, the most common form of the vitamin, is directly involved in photosynthesis. This causes leaf vegetables to be the primary food class that interacts significantly with the anticoagulant pharmaceutical warfarin.

During the first half of the 20th century, it was common for greengrocers to carry small bunches of herbs tied with a string to small green and red peppers, these bundles were called "potherbs."

Lettuce

Lettuce (*Lactuca sativa*) is an annual plant of the aster or sunflower family Asteraceae. It is most often grown as a leaf vegetable, but sometimes for its stem and seeds. Lettuce was first cultivated by the ancient Egyptians who turned it from a weed, whose seeds were used to produce oil, into a plant grown for its leaves. Lettuce spread to the Greeks and Romans, the latter of whom gave it the name "*lactuca*", from which the English "lettuce" is ultimately derived.

By 50 AD, multiple types were described, and lettuce appeared often in medieval writings, including several herbals. The 16th through 18th centuries saw the development of many varieties in Europe, and by the mid-18th century cultivars were described that can still be found in gardens. Europe and North America originally dominated the market for lettuce, but by the late 1900s the consumption of lettuce had spread throughout the world.

Generally grown as a hardy annual, lettuce is easily cultivated, although it requires relatively low temperatures to prevent it from flowering quickly. It can be plagued with numerous nutrient deficiencies, as well as insect and mammal pests and fungal and bacterial diseases. *L. sativa* crosses easily within the species and with some other species within the *Lactuca* genus; although this trait can be a problem to home gardeners who attempt to save seeds, biologists have used it to broaden the gene pool of cultivated lettuce varieties. World production of lettuce and chicory for calendar year 2010 stood at 23,620,000 metric tons (23,250,000 long tons; 26,040,000 short tons), over half of which came from China.

Lettuce is most often used for salads, although it is also seen in other kinds of food, such as soups, sandwiches and wraps; it can also be grilled. One variety, the Woju (′*f*ã,) or asparagus lettuce, is grown for its stems, which are eaten either raw or cooked. Lettuce is a good source of vitamin A and potassium, as well as a minor source for several other vitamins and nutrients.

Despite its beneficial properties, contaminated lettuce is often a source of bacterial, viral and parasitic outbreaks in humans, including *E. coli* and *Salmonella*. In addition to its main use as a leafy green, it has also gathered religious and medicinal significance over centuries of human consumption.

Description

Figure: *Mature lettuce inflorescence in fruit*

Lettuce's native range spreads from the Mediterranean to Siberia, although it has been transported to almost all areas of the world. Plants generally have a height and spread of 6 to 12 inches (15 to 30 cm). The leaves are colourful, mainly in the green and red colour spectrums, with some variegated varieties. There are also a few varieties with yellow, gold or blue-teal leaves. Lettuces have a wide range of shapes and textures, from the dense heads of the iceberg type to the notched, scalloped, frilly or ruffly leaves of leaf varieties. Lettuce plants have a root system that includes a main taproot and smaller secondary roots. Some varieties, especially those found in the US and Western Europe, have long, narrow taproots and a small set of secondary roots. Longer taproots and more extensive secondary systems are found in varieties from Asia.

Depending on the variety and time of year, lettuce generally lives 65–130 days from planting to harvesting. Because lettuce that flowers (through the process known as "bolting") becomes bitter and unsaleable, plants grown for consumption are rarely allowed to grow to maturity. Lettuce flowers more quickly in hot temperatures, while freezing temperatures cause slower growth and sometimes damage to outer leaves. Once plants move past the edible stage, they develop flower stalks up to 3 feet (0.9 m) high with small yellow blossoms. Like other members of the tribe Cichorieae, lettuce inflorescences (also known as flower heads or capitula) are composed of multiple florets, each with a modified calyx called a pappus (which becomes the feathery "parachute" of the fruit), a corolla of five petals fused into a ligule or strap, and the reproductive parts. These include fused anthers that form a tube which surrounds a style and bipartite stigma. As the anthers shed pollen, the style elongates to allow the stigmas, now coated with pollen, to emerge from the tube. The ovaries form compressed, obovate (teardrop-shaped) dry fruits that do not open at maturity, measuring 3 to 4 mm long. The fruits have 5–7 ribs on each side and are tipped by two rows of small white hairs. The pappus remains at the top of each fruit as a dispersal structure. Each fruit contains one seed, which can be white, yellow, gray or brown depending on the variety of lettuce.

The domestication of lettuce over the centuries has resulted in several changes through selective breeding: delayed bolting, larger seeds, larger leaves and heads, better taste and texture, a lower latex content, and different leaf shapes and colours. Work in these areas continues through the present day. Scientific research into the genetic modification of lettuce is ongoing, with over 85 field trials taking place between 1992 and 2005 in the European Union and United States to test modifications allowing greater herbicide tolerance, greater resistance to insects and fungi and slower bolting patterns. However, genetically modified lettuce is not currently used in commercial agriculture.

Cultivation

A hardy annual, some varieties of lettuce can be overwintered even in relatively cold climates under a layer of straw, and older, heirloom varieties are often grown in cold frames. Lettuces meant for the cutting of individual leaves are generally planted straight into the garden in thick rows. Heading varieties of lettuces are commonly started in flats, then transplanted to individual spots, usually 8 to 14 inches (20 to 36 cm) apart, in the garden after developing several leaves. Lettuce spaced further apart receives more sunlight, which improves colour and nutrient quantities in the leaves. Pale to white lettuce, such as the centres in some iceberg lettuce, contain few nutrients.

Lettuce grows best in full sun in loose, nitrogen-rich soils with a pH of between 6.0 and 6.8. Heat generally prompts lettuce to bolt, with most varieties growing poorly above 75 °F (24 °C); cool temperatures prompt better performance, with 60 to 65 °F (16 to 18 °C) being preferred and as low as 45 °F (7 °C) being tolerated. Plants in hot areas that are provided partial shade during the hottest part of the day will bolt more slowly. Temperatures above 80 °F (27 °C) will generally result in poor or non-existent germination of lettuce seeds. After harvest, lettuce lasts the longest when kept at 32 °F (0 °C) and 96 percent humidity. Lettuce quickly degrades when stored with fruit such as apples, pears and bananas that release the ripening agent ethylene gas.

The high water content of lettuce (94.9 percent) creates problems when attempting to preserve the plant – it cannot be successfully frozen, canned or dried and must be eaten fresh.

Lettuce varieties will cross with each other, making spacing of 5 to 20 feet (1.5 to 6.1 m) between varieties necessary to prevent contamination when saving seeds. Lettuce will also cross with *Lactuca serriola* (wild lettuce), with the resulting seeds often producing a plant with tough, bitter leaves. Celtuce, a lettuce variety grown primarily in Asia for its stems, crosses easily with lettuces grown for their leaves.

This propensity for crossing, however, has led to breeding programs using closely related species in *Lactuca*, such as *L. serriola*, *L. saligna*, and *L. virosa*, to broaden the available gene pool. Starting in the 1990s, such programs began to include more distantly related species such as *L. tatarica*. Seeds keep best when stored in cool conditions, and, unless stored cryogenically, remain viable the longest when stored at "4 °F ("20 °C); they are relatively short lived in storage.

At room temperature, lettuce seeds remain viable for only a few months. However, when newly harvested lettuce seed is stored cryogenically, this life increases to a half-life of 500 years for vaporized nitrogen and 3,400 years for liquid nitrogen; this advantage is lost if seeds are not frozen promptly after harvesting.

Cultivars

Figure: *A selection of lettuce cultivars*

There are several types of lettuce, but three (leaf, head and cos or romaine) are the most common. There are seven main cultivar groups of lettuce, each including many varieties:

- Leaf – Also known as looseleaf, cutting or bunching lettuce, this type has loosely bunched leaves and is the most widely planted. It is used mainly for salads.
- Romaine/Cos – Used mainly for salads and sandwiches, this type forms long, upright heads. This is the most often used lettuce in Caesar salads.
- Crisphead – Better known as "iceberg" lettuce, the most popular lettuce in the US, this type is very heat-sensitive and was originally adapted for growth in the northern US. It ships well, but is low in flavour and nutritional content, being composed of even more water than other lettuce types.
- Butterhead – Also known as Boston or Bibb lettuce, this type is a head lettuce with a loose arrangement of leaves, known for its sweet flavour and tender texture.

Figure: Woju *(′fã,), a lettuce variety grown for its stem used in Chinese cooking, sold in a Yangzhou supermarket*

- Summercrisp – Also called Batavian or French Crisp, this lettuce is midway between the crisphead and leaf types. These lettuces tend to be larger, bolt-resistant and well-flavoured.
- Stem – This type is grown for its seedstalk, rather than its leaves, and is used in Asian cooking, primarily Chinese, as well as stewed and creamed dishes.
- Oilseed – This type is grown for its seeds, which are pressed to extract an oil mainly used for cooking. It has few leaves, bolts quickly and produces seeds around 50 percent larger than other types of lettuce.

The butterhead and crisphead types are sometimes known together as "cabbage" lettuce, because their heads are shorter, flatter, and more cabbage-like than romaine lettuces.

Cultivation Problems

Figure: *A lettuce surrounded by weeds, which have crowded it to the point of bolting*

Nutrient deficiencies, including a lack of boron, phosphorus, calcium, lithium, molybdenum or copper, can cause a variety of plant problems that range from malformed plants to a lack of head growth. Many insects are attracted to lettuce, including cutworms, which cut seedlings off at the soil line; wireworms and nematodes, which cause yellow, stunted plants; tarnished plant bugs and aphids, which cause yellow, distorted leaves; leafhoppers, which cause stunted growth and pale leaves; thrips, which turn leaves gray-green or silver; leafminers, which create tunnels within the leaves; flea beetles, which cut small holes in leaves and caterpillars, slugs and snails, which cut large holes in leaves. Mammals, including rabbits and groundhogs, also eat the plants. Lettuce contains several defencive compounds, including sesquiterpene lactones, and other natural phenolics such as flavonol and glycosides, which help to protect it against pests. Certain varieties contain more than others, and some selective breeding and genetic modification studies have focused on using this trait to identify and produce commercial varieties with increased pest resistance.

Lettuce also suffers from several viral diseases, including big vein, which causes yellow, distorted leaves, and mosaic virus, which is spread by aphids and causes stunted plant growth and deformed leaves. Aster yellows are a disease-causing bacteria carried by leafhoppers, which causes deformed leaves. Fungal diseases include powdery mildew and downy mildew, which cause leaves to mold and die and bottom rot, lettuce drop and gray mold, which cause entire plants to rot and collapse. Crowding lettuce tends to attract pests and diseases. Weeds can also be an issue, as cultivated lettuce is generally not competitive with them, especially when directly seeded into the ground. Transplanted lettuce (started in flats and later moved to growing beds) is generally more competitive initially, but can still be crowded later in the season, causing misshapen lettuce and lower yields. Weeds also act as homes for insects and disease and can make harvesting more difficult. Herbicides are often used to control weeds in commercial production. However, this has led to the development of herbicide-resistant weeds and prompted environmental and health concerns.

Production

Lettuce is the only member of the *Lactuca* genus to be grown commercially. The Food and Agriculture Organisation of the United Nations (FAO) reports that world production of lettuce and chicory (the two crops are combined by the FAO for reporting purposes) for calendar year 2010 was 23,622,366 metric tons (23,249,287 long tons; 26,039,201 short tons). This came primarily from China (53 percent), the US (17 percent) and India (4 percent). Although China is the top world producer of lettuce, the majority of the crop is consumed domestically. Spain is the world's largest exporter of lettuce, with the US ranking second.

Western Europe and North America were the original major markets for large-scale lettuce production. By the late 1900s, Asia, South America, Australia and Africa became more substantial markets. Different locations tended to prefer different types of lettuce, with butterhead prevailing in northern Europe and Great Britain, romaine in the Mediterranean and stem lettuce in China and Egypt. By the late 20th century, the preferred types began to change, with crisphead, especially iceberg, lettuce becoming the dominant type in northern Europe and Great Britain and more popular in western Europe.

In the US, no one type predominated until the early 20th century, when crisphead lettuces began gaining popularity. After the 1940s, with the development of iceberg lettuce, 95 percent of the lettuce grown and consumed in the US was crisphead lettuce. By the end of the century, other types began to regain popularity and eventually made up over 30 percent of production. Stem lettuce was first developed in China, and remains primarily cultivated in that country.

Lettuce production methods, including all of the processes from growing to sales, have become much larger in scale during the 20th century. The majority of agricultural production is done with the application of large amounts of chemicals, including fertilizers and pesticides, but organic production makes up a growing percentage of the market – a trend that began with small growers but moved to a more industrial scale.

More non-heading types, mostly leaf and romaine lettuces, are also being grown. In the first years of the 21st century, bagged salad products began to hold a growing portion of the lettuce market, especially in the US. Processed from what was previously waste lettuce not considered acceptable for the fresh market, these products are packaged in a manner that makes them last longer than standard head lettuce after harvest. As of 2007, 70 percent of the lettuce production in the US came from California; in that country it ranks third in produce consumption behind tomatoes and oranges.

Spinach

Spinach (*Spinacia oleracea*) is an edible flowering plant in the family of Amaranthaceae. It is native to central and southwestern Asia. It is an annual plant (rarely biennial), which grows to a height of up to 30 cm. Spinach may survive over winter in temperate regions. The leaves are alternate, simple, ovate to triangular-based, very variable in size from about 2–30 cm long and 1–15 cm broad, with larger leaves at the base of the plant and small leaves higher on the flowering stem. The flowers are inconspicuous, yellow-green, 3–4 mm diameter, maturing into a small, hard, dry, lumpy fruit cluster 5–10 mm across containing several seeds.

Common spinach, *Spinacia oleracea*, was long considered to be in the Chenopodiaceae family, but in 2003, the Chenopodiaceae family was combined with the Amaranthaceae family under the family name 'Amaranthaceae' in the order Caryophyllales. Within the Amaranthaceae family, Amaranthoideae and Chenopodioideae are now subfamilies, for the amaranths and the chenopods, respectively.

Types of Spinach

A distinction can be made between older varieties of spinach and more modern ones. Older varieties tend to bolt too early in warm conditions. Newer varieties tend to grow more rapidly, but have less of an inclination to run up to seed. The older varieties have narrower leaves and tend to have a stronger and more bitter taste. Most newer varieties have broader leaves and round seeds.

The three basic types of spinach are:

- Savoy has dark green, crinkly and curly leaves. It is the type sold in fresh bunches in most supermarkets in the United States. One heirloom variety of savoy is *Bloomsdale*, which is somewhat resistant to bolting. Other common heirloom varieties are *Merlo Nero* (a mild variety from Italy) and *Viroflay* (a very large spinach with great yields).
- Flat- or smooth-leaf spinach has broad, smooth leaves that are easier to clean than Savoy. This type is often grown for canned and frozen spinach, as well as soups, baby foods, and processed foods. *Giant Noble* is an example variety.
- Semi-savoy is a hybrid variety with slightly crinkled leaves. It has the same texture as Savoy, but it is not as difficult to clean. It is grown for both fresh market and processing. *Tyee Hybrid* is a common semi-savoy.

Production, Marketing and Storage

Spinach is sold loose, bunched, packaged fresh in bags, canned, or frozen. Fresh spinach loses much of its nutritional value with storage of more than a few days. While refrigeration slows this effect to about eight days, spinach will lose most of its folate and carotenoid content, so for longer storage, it is blanched and frozen, cooked and frozen, or canned. Storage in the freezer can be for up to eight months.

Dealing with Spinach Pests and Diseases

Easy to grow and quick to harvest, spinach is one of the mainstays of the vegetable garden. It grows best in the cool part of the year, but with bolt-resistant varieties and a little shade, you can grow spinach in summer, too. This nutritious vegetable is tasty when served raw or cooked, but unfortunately, pesky insects love it just as much.

Common Spinach Pests

There are a number of insects that feast on spinach plants. However, the most common spinach pests that affect these plants include the following:

- Cutworms and wireworms – Cutworms cut young seedlings off at ground level and wireworms feed on the foliage and roots. Older transplants are less appealing to these insects than tender seedlings. To trap wireworms, plant full grown carrots in the garden every 2 ½ to 3 feet. Pull up the carrots every two or three days and remove the trapped wireworms, then replace the carrot in the garden. Cutworms respond to Bacillus thuringiensis (Bt) and spinosad spray.

- Flea beetles – Flea beetles feed on young foliage. The damage consists of a number of small holes, leaving the leaf looking as though it's been hit by a blast from a shotgun.

  The leaves sometimes have bleached and pitted areas as well. The insects are so tiny that you may never see them. Use reflective mulch or lay sheets of aluminum foil under the plants. Carbaryl and pyrethrum insecticides sometimes help reduce serious infestations.
- Slugs and snails – Slugs and snails also leave holes in spinach leaves. You can tell the difference by the size of the hole—slug and snail holes are much larger—and by the slime trail slugs and snails leave behind. Baits and traps are the best way to control these insects.
- Aphids – Aphids are probably the most common of spinach pests. Natural enemies generally keep them in check. If you need more help, use insecticidal soaps or neem oil.
- Leaf miners – Leaf miners leave meandering tan trails on the leaves. Since they are feeding inside the leaves, contact insecticides aren't effective. Clip off infested leaves and destroy them before the larvae have a chance to mature.

Spinach Diseases

Just as with pests, other issues with spinach can also pop up in the garden. Of those most often seen are spinach diseases, which include:

- Damping off – Damping off disease causes seedlings to fall over and die soon after they emerge. Plant quality seeds and avoid overwatering to help prevent the disease. Process compost thoroughly in a hot pile before adding it to the garden soil.
- Downy mildew – Downy mildew causes yellow or light green spots on the upper surface of leaves with white fungus on the lower surface. There is no cure, and the best course of action is to remove infected plants. Preventative measures include spacing plants at the recommended distance to allow good air circulation and applying water directly to the soil to keep the foliage dry. Avoid planting spinach next year if you have problems with downy mildew this year. This gives the disease spores a chance to die out.
- Viruses – Viruses that infect spinach are often spread by insects, so control insect infestations as much as possible. There is no

cure for infected plants. Pull up and destroy the plants to prevent the spread of the virus.

Tackling Common Spinach Problems

Common spinach problems and issues with spinach are sometimes caused by the way we grow the plant and environmental conditions. Spinach grows and tastes best in cool weather. In hot weather, seeds are slow to germinate and may not germinate at all. Heat also causes the plants to bolt (go to seed) quickly, ruining the flavour of the crop.

Plant the seeds in late winter or early spring for an early crop or in late summer or early fall for a fall or winter crop. If you want to try growing spinach in summer, plant it where it will receive a half day of shade.

Salad Crops

Salad is a popular ready-to-eat dish often containing leafy vegetables, usually served chilled or at a moderate temperature and often served with a sauce or dressing. Salads may also contain ingredients such as fruit, grain, meat, seafood and sweets. Though many salads use raw ingredients, some use cooked ingredients.

Lettuce, endive, celery, and parsley are the most popular salad vegetables. There are three types of lettuce: looseleaf, head or semi-head, and upright. Leaf lettuce grows quickly and is the easiest type of lettuce to grow. Endive is available in finely curled and broad-leaved varieties. The broad-leaved variety is often called escarole. Celery is not usually grown in the home garden. It requires a long growing season, and it does not grow well in hot, dry summer weather. Parsley is used for garnishing and flavouring. A few plants will produce enough for the average family.

The salad crops provide vitamins as well as needed bulk. They are usually eaten without cooking.

Preparing the Soil and Fertilizing

Proper soil preparation is very important for success with the salad crops. They have small root systems and are poor foragers, so there needs to be a good supply of nutrients in the surface soil. The soil should be well drained and moisture retentive. Deeply plow or till the soil, then pulverize and smooth the surface. Incorporate well-rotted manure, other organic matter, compost, and a complete fertilizer such as a 10-10-10, at the rate of 2 pounds per 100 square feet into the soil before planting. In the midseason you should apply a side dressing of the same fertilizer at a rate of 1 pound for each 25 feet of row. Do not use a combination of fertilizer and herbicide on vegetables. They contain weed killers that will kill vegetable plants.

The optimum pH range for salad crops is between 5.5 and 6.5. Liming will raise the pH of acid soils. Use a soil test to determine if any special fertilization or liming is required.

Planting and Transplanting

Seed leaf lettuce directly in the garden April 20 for a spring crop, and again August 1 for a fall crop. Space plants 2 inches apart. Head lettuce is more difficult to grow and is not recommended for inexperienced gardeners. For best results, start seeds indoors April 1 and transplant to the garden May 10. Space head lettuce 10 inches apart.

Seed endive directly in the garden April 20 for a spring crop, and again July 1 for a fall crop. Space plants 10 inches apart. Blanching endive will produce white leaves in the heart of the plant and prevent bitterness. To blanch a plant, tie the outer leaves up around the heart of the plant. Blanching takes about two weeks.

Celery requires a very long growing season and must be started indoors. Start seeds February 1 and transplant to the garden May 10. Space plants 15 inches apart. Thinly seed parsley directly in the garden May 10. After growth begins, thin so the plants are 4 to 6 inches apart. Plants may be started indoors.

Note: Planting dates given are for the Minneapolis-St. Paul area. Plant one week earlier or later for each 100 miles south or north.

Watering

Watering is essential to successfully grow the salad crop. Their small, shallow root system cannot tolerate dry weather. They need at least 1 inch of water from rainfall or irrigation each week during the growing season. Always soak the soil thoroughly when watering. There is little or no value in a light watering that only wets the surface of the soil. Lettuce, endive, and celery will be crisper if they are watered several days prior to harvesting. The leaves of lettuce and endive will be tough and the edges will turn brown if they do not receive enough water.

Controlling Weeds

The salad crops are slow growing and they have small root systems, so they cannot successfully compete with weeds. Frequent, shallow cultivation will control the weeds and keep the surface of the soil loose. The roots of the salad crops are very close to the surface, so it is important to not cultivate too deeply. Cultivate just deeply enough to cut the weeds off below the soil surface. Do not overcultivate. Unnecessary cultivation may dry out the surface of the soil.

Harvesting

Leaf lettuce can be harvested as soon as the leaves become large enough to use. You can remove the whole plant by cutting it off at or just below the surface of the soil. However, by removing only the outer leaves each time, the plant will continue to produce for a long time. Head lettuce can be harvested before the head is well formed. Harvest by cutting the head off at or just below the surface of the soil. For best quality, lettuce should be harvested before the hot, dry weather causes it to become bitter.

Harvest endive when the plant reaches a usable size and after it has been blanched. To harvest, cut the plant off at or just below the surface of the soil. Harvest celery as soon as it reaches a usable size. It does not have to be full grown. To harvest, cut the plant off below the surface of the soil. Celery harvested in hot, dry weather may be tough, stringy, and bitter.

To harvest parsley, pick only a few leaves from a plant at a time. Plants will produce for several weeks. In late fall you can dig up parsley plants and bring them indoors to produce fresh parsley during the winter.

Problems

Dwarfed, yellow plants may be caused by yellows disease and spread by leafhoppers. Control by controlling leafhoppers. Do not apply poisonous insecticides to plant parts that are to be eaten.

Holes in the leaves are caused by snails and slugs. Wet weather usually worsens the condition.

Tomato

The tomato is the edible, often red fruit/berry of the nightshade *Solanum lycopersicum*, commonly known as a tomato plant. The species originated in the South American Andes and its use as a food originated in Mexico, and spread throughout the world following the Spanish colonization of the Americas. Its many varieties are now widely grown, sometimes in greenhouses in cooler climates.

The tomato is consumed in diverse ways, including raw, as an ingredient in many dishes, sauces, salads, and drinks. While it is botanically a fruit, it is considered a vegetable for culinary purposes (as well as under U.S. customs regulations), which has caused some confusion. The fruit is rich in lycopene, which may have beneficial health effects.

The tomato belongs to the nightshade family, *Solanaceae*. The plants typically grow to 1–3 metres (3–10 ft) in height and have a weak stem that often sprawls over the ground and vines over other plants. It is a perennial in its native habitat, although often grown outdoors in temperate climates as an annual. An average common tomato weighs approximately 100 grams (4 oz).

The poor taste and lack of sugar in modern garden and commercial tomato varieties resulted from breeding tomatoes to ripen uniformly red. This change occurred after discovery of a variety in the mid 20th century that ripened uniformly. This was widely crossbred to produce red fruit without the typical green ring around the stem on uncrossbred varieties. Prior to general introduction of this trait, most tomatoes produced more sugar during ripening, and were sweeter and more flavourful.

Cultivation

The tomato is now grown worldwide for its edible fruits, with thousands of cultivars having been selected with varying fruit types, and for optimum growth in differing growing conditions. Cultivated tomatoes vary in size, from tomberries, about 5 mm in diameter, through cherry tomatoes, about the same 1–2 cm (0.4–0.8 in) size as the wild tomato, up to beefsteak tomatoes 10 cm (4 in) or more in diameter. The most widely grown commercial tomatoes tend to be in the 5–6 cm (2.0–2.4 in) diameter range.

Figure: *Tomato plants 7 days after planting*

Figure: *27 days after planting*

Figure: *52 day old plant, first fruits*

Most cultivars produce red fruit, but a number of cultivars with yellow, orange, pink, purple, green, black, or white fruit are also available. Multicoloured and striped fruit can also be quite striking. Tomatoes grown for canning and sauces are often elongated, 7–9 cm (3–4 in) long and 4–5 cm (1.6–2.0 in) diameter; they are known as plum tomatoes, and have a lower water content. Roma-type tomatoes are important cultivars in the Sacramento Valley.

Tomatoes are one of the most common garden fruits in the United States and, along with zucchini, have a reputation for outproducing the needs of the grower.

Quite a few seed merchants and banks provide a large selection of heirloom seeds. The definition of an heirloom tomato is vague, but unlike commercial hybrids, all are self-pollinators that have bred true for 40 years or more. About 150 million tons of tomatoes were produced in the world in 2009. China, the largest producer, accounted for about one quarter of the global output, followed by United States and India. For one variety, plum or processing tomatoes, California accounts for 90% of U.S. production and 35% of world production.

Varieties

There are around 7500 tomato varieties grown for various purposes. Heirloom tomatoes are becoming increasingly popular, particularly among home gardeners and organic producers, since they tend to produce more interesting and flavourful crops at the cost of disease resistance and productivity. In 1973, Israeli scientists developed the world's first long shelf-life commercial tomato varieties.

Hybrid plants remain common, since they tend to be heavier producers, and sometimes combine unusual characteristics of heirloom tomatoes with the ruggedness of conventional commercial tomatoes.

Tomato varieties are roughly divided into several categories, based mostly on shape and size.

- "Slicing" or "globe" tomatoes are the usual tomatoes of commerce, used for a wide variety of processing and fresh eating.
- Beefsteak tomatoes are large tomatoes often used for sandwiches and similar applications. Their kidney-bean shape, thinner skin, and shorter shelf life makes commercial use impractical.
- Oxheart tomatoes can range in size up to beefsteaks, and are shaped like large strawberries.
- Plum tomatoes, or paste tomatoes (including pear tomatoes), are bred with a higher solids content for use in tomato sauce and paste, and are usually oblong.
- Pear tomatoes are pear-shaped, and are based upon the San Marzano types for a richer gourmet paste.
- Cherry tomatoes are small and round, often sweet tomatoes generally eaten whole in salads.
- Grape tomatoes, a more recent introduction, are smaller and oblong, a variation on plum tomatoes, and used in salads.
- Campari tomatoes are also sweet and noted for their juiciness, low acidity, and lack of mealiness. They are bigger than cherry tomatoes, but are smaller than plum tomatoes.

Early tomatoes and cool-summer tomatoes bear fruit even where nights are cool, which usually discourages fruit set. There are also varieties high in beta carotenes and vitamin A, hollow tomatoes and tomatoes that keep for months in storage.

Tomatoes are also commonly classified as determinate or indeterminate. Determinate, or bush, types bear a full crop all at once and top off at a specific height; they are often good choices for container growing. Determinate types are preferred by commercial growers who wish to harvest a whole field at one time, or home growers interested in canning. Indeterminate varieties develop into vines that never top off and continue producing until killed by frost. They are preferred by home growers and local-market farmers who want ripe fruit throughout the season. As an intermediate form, there are plants sometimes known as vigorous determinate or semideterminate; these top off like determinates, but produce a second crop after the initial crop. The majority of heirloom tomatoes are indeterminate, although some determinate heirlooms exist.

Most modern tomato cultivars are smooth surfaced, but some older tomato cultivars and most modern beefsteaks often show pronounced ribbing, a feature that may have been common to virtually all pre-Columbian cultivars. While virtually all commercial tomato varieties are red, some cultivars – especially heirlooms – produce fruit in other colours, including green, yellow, orange, pink, black, brown, ivory, white, and purple. Such fruits are not widely available in grocery stores, nor are their seedlings available in typical nurseries, but they can be bought as seed. Less common variations include fruit with stripes (Green Zebra), fuzzy skin on the fruit (Fuzzy Peach, Red Boar), multiple colours (Hillbilly, Burracker's Favourite, Lucky Cross), etc.

There is also a considerable gap between commercial and home-gardener cultivars. Home cultivars are often bred for flavour to the exclusion of all other qualities, while commercial cultivars are bred for factors like consistent size and shape, disease and pest resistance, suitability for mechanized picking and shipping, and ability to ripen after picking.

Tomatoes grow well with seven hours of sunlight a day. A fertilizer with an NPK ratio of 5-10-10 is often sold as tomato fertilizer or vegetable fertilizer, although manure and compost are also used.

Diseases and Pests

Tomato cultivars vary widely in their resistance to disease. Modern hybrids focus on improving disease resistance over the heirloom plants. One common tomato disease is tobacco mosaic virus. Handling cigarettes and other infected tobacco products can transmit the virus to tomato plants. Various forms of mildew and blight are also common tomato afflictions, which is why tomato cultivars are often marked with a combination of letters that refer to specific disease resistance. The most common letters are: V – *verticillium* wilt, F – *fusarium* wilt strain I, FF – *fusarium* wilt strain I and II, N – *nematodes*, T – *tobacco mosaic virus*, and A – *alternaria.*

Figure: *Tomato fruitworm feeding on unripe tomato*

Another particularly dreaded disease is curly top, carried by the beet leafhopper, which interrupts the lifecycle, ruining a nightshade plant as a crop. As the name implies, it has the symptom of making the top leaves of the plant wrinkle up and grow abnormally. Some common tomato pests are stink bugs, cutworms, tomato hornworms and tobacco hornworms, aphids, cabbage loopers, whiteflies, tomato fruitworms, flea beetles, red spider mite, slugs, and Colorado potato beetles.

Tomato plants produce the plant peptide hormone systemin after an insect attack. Systemin activates defencive mechanisms, such as the production of protease inhibitors to slow the growth of insects. The hormone was first identified in tomatoes, but similar proteins have been identified in other species since.

How to Identify Tomato Pests and Control Them

Left unchecked, tomato pests can inflict lots of damage to your tomato plants and even destroy them.

But there's good news. If you keep a close watch, you can identify pests as soon as they begin their dirty work and treat them before things get out of hand.

Aphids: These are those dense clusters of tiny insects you may see on the stems or new growth of your tomato plants. While small numbers are not a problem, large infestations can gradually injure or even kill plants. Pinch off foliage where aphids are densely concentrated, and throw these discarded bits into the garbage, not on the ground. If the problem then seems manageable, release beneficial insects such as ladybugs or lacewings; if it doesn't, go for the insecticidal soap or organic sprays.

Cutworms: These are the tiny caterpillars that feed on young plant stems at night, frequently felling seedlings by eating right through them at ground level. Prevent damage by placing collars around seedlings. You can make these of paper, cardboard, aluminum foil, or an aluminum pie plate about ten inches long and four high, bent to form a circle or cylinder and stapled. Sink the collars about an inch into soil around individual seedlings, letting three inches show above the ground to deter high-climbers.

Flea Beetles: A potentially devastating visitor, the flea beetle (so-named because it resembles and jumps like a flea) attacks from both sides: adults eat foliage, leaving numerous small holes, while larvae feed on roots. They're not picky, these beetles; they'll go for corn, cabbage, lettuce, and all members of the Solanaceae family: peppers, eggplant, potatoes, and tomatoes. Unless levels are very high, damage can be minimized and controlled by using preventative measures. Clear away or plow under weeds and debris, in which adults over-winter, and place yellow sticky-traps to monitor levels and capture adults. Young plants are more vulnerable to damage, so cover them with row covers. Dusting plants with diatomaceous earth (a chalky stone composed of marine fossils, ground to powder) helps control adults feeding on foliage. To attack the insect more directly, introduce beneficial nematodes into your soil to feed on the larvae and pupae. In cases of high infestation and serious damage, botanical insecticides such as rotenone or pyrethrin can be used.

Hornworms: These caterpillars are so big (three inches long or more) that it would seem to be easy to control them just by picking them off. And so it is, sometimes. The problem is that their pale green colour provides excellent camouflage, and the nymph and larval stages are far smaller and less obvious. If there are only a few, picking them off works well. (One site suggests spraying the plant with water, causing the caterpillars to, and I quote, "thrash around," giving themselves away.) If there are more than a few, other measures may be called for. One of these is Bt, or *Bacillus thuringiensis*, an organic treatment that can control numerous other problems as well.

Nematodes: This is one of the most dreaded tomato problems. Actually, almost 20,000 different species of nematode have been identified, and billions of these usually microscopic worms occupy each acre of fertile earth, so it is fortunate that only a few cause gardening problems. Some, insect pathogenic nematodes, can actually help control other gardening pests such as fungus gnats or flea beetles. But when a gardening friend says in a voice of doom, "I've got nematodes," he generally means one thing: root-knot nematodes. This particular species invades various crops, causing bumps or galls that interfere with the plant's ability to take up nutrients and to perform photosynthesis. They're most common in warmer areas with short winters.

Unfortunately, controlling nematodes is not easy. Since they take several seasons to get established, rotating crops denies this pest the chance to get entrenched. It's crucial, though, that you follow tomatoes with crops that are not vulnerable to the same problem! Members of the same family are of course taboo; this includes peppers, eggplant, and potatoes. However, less likely crops are also vulnerable; these include okra and cotton, in the south, and peas, squash, beets, and numerous others anywhere. If you suspect nematodes — if you ever pull a plant that has odd-looking lumpy growths on its roots — have your county extension agent take a look at it, and get advice about crop rotation in your area.

Completely sterilizing the soil is one option on small plots, but it's toxic and sometimes expensive. It also means that you've killed off all the beneficial organisms in the soil as well as the troublesome ones, so it's particularly important to follow such treatment with a big infusion of clean compost. It would also be best to add earthworms, beneficial nematodes, and an assortment of micro-organisms as well, since doing so will restore the soil to full health and make it less vulnerable to further incursions by nematodes.

While eliminating nematodes is extremely difficult, it is possible to limit their damage by using resistant varieties, marked N. Doing so doesn't kill the pests, but it does keep them and their effects under control.

Whiteflies: These tiny insects feed on plant juices, leaving behind a sticky residue or 'honeydew,' which can become a host for sooty mold. Rustle the leaves of infested plants, and clouds of these insects will rise. If you have a serious problem, you may be tempted to reach for a conventional insecticide, but don't bother, as whiteflies have developed resistance to many. The best bet is a horticultural oil, which effectively smothers all stages of this insect. To deal with lower levels, place yellow sticky traps to monitor and suppress infestations. Hosing down plants can be surprisingly effective, especially if you use a bug-blaster, a hose attachment designed to produce an intense multi-directional spray that easily reaches the undersides of leaves. Another tactic is to release natural predators such as ladybugs, lacewings, or whitefly parasites. If the situation is out of control, insecticidal soaps and botanical insecticides and oils can bring populations down to manageable levels, at which point natural predators can maintain them.

Common Tomato Pest Symptoms

Monitor tomato plants daily. Check leaves, stems and fruit for these symptoms that come from nasty critters on the prowl.

On Leaves and Stems

Symptom	*Pest*
chewed stem	tomato cutworm (early in the season)
defoliation	tomato hornworm, tobacco hornworm
holes in leaves	flea beetles
yellowing, curled leaves	aphids
purpling veins in leaves	psyllids
tunnels or zigzag patterns in leaves	leaf miners
hole in stem	stalk borer
honeydew (white, sticky residue)	aphids, whiteflies
webbing on leaf undersides	spider mites

On Fruit

Symptom	*Pest*
dark pinpricks, holes in fruit	tomato fruit worm, stink bugs
light or discoloured patches on fruit	stink bugs
holes in fruit	slugs

How to Treat Pests on Tomatoes

Here's an overview of the most common treatment methods for tomato pests. Choose the best treatment prescribed for each one.

- Spray with water. Use a hose to spray plants with a strong jet water stream. This causes insects (especially smaller varieties) to dislodge and move on. Spray plants several days in a row to eradicate as many generations of fast-multiplying insects as possible.
- Handpick insects. Remove insects by hand. Use gloves. Destroy pests after removal by drowning them in a bucket of soapy water or crushing them with your foot. While handpicking is not practical for very small pests, it works well with larger ones like the tomato hornworm and the tobacco hornworm.
- Apply insecticidal soap . Potassium salts in insecticidal soap's fatty acids remove an insect's protective waxes and cause disruption of insect cell membranes. They are effective when applied directly to the insect on the plant. Mix with water to

produce a 2-3% dilution. One advantage of using insecticidal soap is less residual effects on other organisms. Moreover, insecticidal soaps are organic. One disadvantage is that certain plants may be sensitive to soaps, causing leaf burn. Insecticidal soaps are particularly effective on smaller pests such as aphids, pysillids, and spider mites.

- Apply horticultural oils. Plant- or petroleum-based oils are combined with water to produce horticultural sprays. Neem oil, for instance, is derived from seed extracts of the neem plant. Oil-based sprays block an insect's air holes, interfere with an insect's metabolism, disrupt feeding, or inhibit insect growth. Like insecticidal soaps, horticultural oils can cause plant injury if not properly diluted.
- Apply insecticides. If pests continue to attack plants after you've tried other treatment methods, you may need to resort to insecticides like Sevin (available in granules , concentrate , or ready to use) or Eight(Permethrin) - available in dust , concentrate , and ready to spray . But keep in mind that by doing so, you may kill helpful insects. Follow product instructions. Treat for 2-4 weeks with a garden sprayer.

Cucumbers

*Cucumber (*Cucumis sativus*) is a widely cultivated plant in the gourd family Cucurbitaceae. It is a creeping vine that bears cylindrical fruits that are used as culinary vegetables. There are three main varieties of cucumber:* slicing, pickling, *and* burpless. *Within these varieties, several different cultivars have emerged. The cucumber is originally from Southern Asia, but now grows on most continents. Many different varieties are traded on the global market.*

Description

The cucumber is a creeping vine that roots in the ground and grows up trellises or other supporting frames, wrapping around supports with thin, spiraling tendrils. The plant has large leaves that form a canopy over the fruit. The fruit of the cucumber is roughly cylindrical, elongated with tapered ends, and may be as large as 60 centimetres (24 in) long and 10 centimetres (3.9 in) in diameter. Having an enclosed seed and developing from a flower, botanically speaking, cucumbers are classified as accessory fruits. Much like tomatoes and squash they are often also perceived, prepared and eaten as vegetables. Cucumbers are usually more than 90% water.

Flowering and Pollination

A few cultivars of cucumber are parthenocarpic, the blossoms creating seedless fruit without pollination. Pollination for these cultivars degrades the quality.

In the United States, these are usually grown in greenhouses, where bees are excluded. In Europe, they are grown outdoors in some regions, and bees are excluded from these areas.

Most cucumber cultivars, however, are seeded and require pollination. Thousands of hives of honey bees are annually carried to cucumber fields just before bloom for this purpose. Cucumbers may also be pollinated by bumblebees and several other bee species.

Most cucumbers that require pollination are self-incompatible, so pollen from a different plant is required to form seeds and fruit. Some self-compatible cultivars exist that are related to the 'Lemon' cultivar.

Symptoms of inadequate pollination include fruit abortion and misshapen fruit. Partially pollinated flowers may develop fruit that are green and develop normally near the stem end, but are pale yellow and withered at the blossom end.

Traditional cultivars produce male blossoms first, then female, in about equivalent numbers. Newer gynoecious hybrid cultivars produce almost all female blossoms. They may have a pollenizer cultivar interplanted, and the number of beehives per unit area is increased, but temperature changes induce male flowers even on these plants, which may be sufficient for pollination to occur.

Insecticide applications for insect pests must be done very carefully to avoid killing off the insect pollinators.

***Figure:** Cucumber, whole and in longitudinal section*

Genome

In 2009, an international team of researchers announced they had sequenced the cucumber genome.

Production

According to the Food and Agriculture Organisation of the United Nations, China produced at least 60% of the global output of cucumbers in 2005, followed at a distance by Turkey, Russia, Iran and the United States.

Cultivation

History: The cucumber originated in India, where a great many varieties have been observed, from *Cucumis hystrix.* It has been cultivated for at least 3,000 years, and was probably introduced to other parts of Europe by the Greeks or Romans. Records of cucumber cultivation appear in France in the 9th century, England in the 14th century, and in North America by the mid-16th century.

Earliest Cultivation

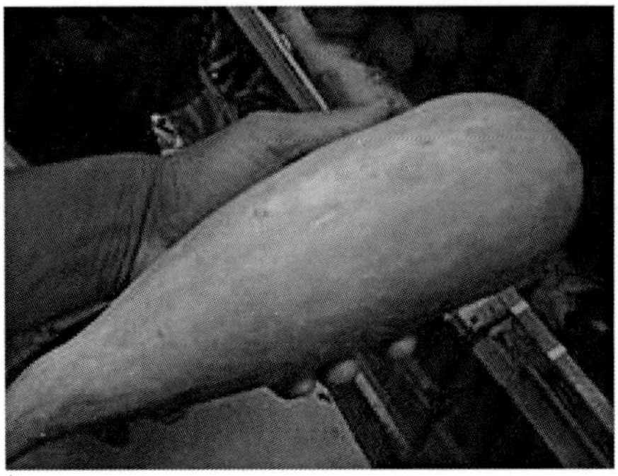

Figure: *An Indian yellow cucumber*

The cucumber is listed among the foods of ancient Ur, and the legend of Gilgamesh describes people eating cucumbers. Some sources also state it was produced in ancient Thrace, and it is certainly part of modern cuisine in Bulgaria and Turkey, parts of which make up that ancient state. Cucumbers are mentioned in the Bible as one of the foods eaten by the Israelites in Egypt. From India, it spread to Greece (where it was called "óßêõïí", *síkyon*) and Italy (where the Romans were especially fond of the crop), and later into China.

Robert Daniel, in discussing an ostracon dated to the second half of the third century AD, has suggested identifying an otherwise unknown word, ïëãéôôá, with the Arabic *al-qitta'*, the common word for cucumber.

According to Pliny the Elder (*The Natural History*, Book XIX, Chapter 23), the Ancient Greeks grew cucumbers, and there were different varieties in Italy, Africa, and Moesia.

Roman Empire

According to Pliny, the Emperor Tiberius had the cucumber on his table daily during summer and winter. The Romans reportedly used artificial methods (similar to the greenhouse system) of growing to have it available for his table every day of the year. "Indeed, he was never without it; for he had raised beds made in frames upon wheels, by means of which the cucumbers were moved and exposed to the full heat of the sun; while, in winter, they were withdrawn, and placed under the protection of frames glazed with mirrorstone."

Reportedly, they were also cultivated in cucumber houses glazed with oiled cloth known as "specularia".

Pliny the Elder describes the Italian fruit as very small, probably like a gherkin, describing it as a wild cucumber considerably smaller than the cultivated one. Pliny also describes the preparation of a medication known as *elaterium*, though some scholars believe he was referring to *Ecballium elaterium*, known in pre-Linnean times as "Cucumis silvestris" or "Cucumis asininus" ("wild cucumber" or "donkey cucumber"), a species different from the common cucumber. Pliny also writes about several other varieties of cucumber, including the cultivated cucumber, and remedies from the different types (9 from the cultivated, 5 from the "anguine", and 26 from the "wild"). The Romans are reported to have used cucumbers to treat scorpion bites, bad eyesight, and to scare away mice. Wives wishing for children wore them around their waists. They were also carried by midwives, and thrown away when the child was born.

Middle Ages

Charlemagne had cucumbers grown in his gardens in the 8th/9th century. They were reportedly introduced into England in the early 14th century, lost, then reintroduced approximately 250 years later. The Spaniards (through the Italian Christopher Columbus) brought cucumbers to Haiti in 1494. In 1535, Jacques Cartier, a French explorer, found "very great cucumbers" grown on the site of what is now Montreal.

Post-Enlightenment

Throughout the 16th century, European trappers, traders, bison hunters, and explorers bartered for the products of American Indian agriculture. The tribes of the Great Plains and the Rocky Mountains learned from the Spanish how to grow European crops. The best farmers on the Great Plains were the Mandan Indians in what is now North and South Dakota. They obtained cucumbers and watermelons from the Spanish, and added them to the crops they were already growing, including several varieties of corn and beans, pumpkins, squash, and gourd plants. The Iroquois were also growing them when the first Europeans visited them.

In 1630, the Reverend Francis Higginson produced a book called *New England's Plantation* in which, describing a garden on Conant's Island in Boston Harbor known as *The Governor's Garden*, he states: "The countrie aboundeth naturally with store of roots of great variety and good to eat. Our turnips, parsnips, and carrots are here both bigger and sweeter than is ordinary to be found in England. Here are store of pompions, cowcumbers, and other things of that nature which I know not…"

William Wood also published in 1633's New England Prospect (published in England) observations he made in 1629 in America: "The ground affords very good itchen gardens, for Turneps, Parsnips, Carrots, Radishes, and Pompions, Muskmillons, Isquoter-squashes, coucumbars, Onions, and whatever grows well in England grows as well there, many things being better and larger."

In the later 17th century, a prejudice developed against uncooked vegetables and fruits. A number of articles in contemporary health publications stated that uncooked plants brought on summer diseases and should be forbidden to children.

The cucumber kept this vile reputation for an inordinate period of time: "fit only for consumption by cows," which some believe is why it gained the name, *cowcumber*.

A copper etching made by Maddalena Bouchard between 1772 and 1793 shows this plant to have smaller, almost bean-shaped fruits, and small yellow flowers. The small form of the cucumber is figured in Herbals of the 16th century, but states, "If hung in a tube while in blossom, the Cucumber will grow to a most surprising length."

Samuel Pepys wrote in his diary on August 22, 1663: "this day Sir W. Batten tells me that Mr. Newburne is dead of eating cowcumbers, of which the other day I heard of another, I think." In "The Greenstone Door", William Satchell notes that "Te Moanaroa was dead – of a surfeit of cucumbers…", having eaten four of the "prickly" melons. (Chapter XX, The Storm Cloud).

Varieties

In human cultivation, the varieties of cucumbers are classified into three main varieties: "slicing", "pickling", and "burpless".

Slicing

Cucumbers grown to eat fresh are called slicing cucumbers. They are mainly eaten in the unripe green form, since the ripe yellow form normally becomes bitter and sour. Slicers grown commercially for the North American market are generally longer, smoother, more uniform in colour, and have a much tougher skin. Slicers in other countries are smaller and have a thinner, more delicate skin. Smaller slicing cucumbers can also be pickled.

Figure: *Slicing cucumbers*

Pickling

Figure: *Pickling cucumbers*

Cucumbers can be pickled for flavour and longer shelf-life. Although any cucumber can be pickled, commercial pickles are made from cucumbers specially bred for uniformity of length-to-diameter ratio and lack of voids in the flesh. Those cucumbers intended for pickling, called picklers, grow to about 7 cm (3 in) to 10 cm (4 in) long and 2.5 cm (1 in) wide. Compared to slicers, picklers tend to be shorter, thicker, less regularly shaped, and have bumpy skin with tiny white or black-dotted spines. They are never waxed. Colour can vary from creamy yellow to pale or dark green. Pickling cucumbers are sometimes sold fresh as "Kirby" or "Liberty" cucumbers. The pickling process removes or degrades much of the nutrient content, especially that of vitamin C. Pickled cucumbers are soaked in brine or a combination of vinegar and brine, although not vinegar alone, often along with various spices. Pickled cucumbers are called "pickles" in the US or "gherkins" or "wallies" in the UK, the latter name being more common in the north of England, where it refers to the large vinegar-pickled cucumbers commonly sold in fish and chip shops. (Although the gherkin is of the same species as the cucumber, it is of a completely different cultivar.)

Burpless

Figure: *Isfahan burpless cucumber originally from Iran*

Burpless cucumbers are sweeter and have a thinner skin than other varieties of cucumber, and are reputed to be easy to digest and to have a pleasant taste. They can grow as long as 2 feet (0.61 m). They are nearly seedless, and have a delicate skin. Most commonly grown in greenhouses, these parthenocarpic cucumbers are often found in grocery markets, shrink-wrapped in plastic. They are sometimes marketed as seedless or burpless, because the seeds and skin of other varieties of cucumbers are said to give some people gas.

Several varietals exist and are sold commercially:

Figure: *'Dosakai' is a round, yellow, cucumber seen at a market in Guntur, India*

- Lebanese cucumbers are small, smooth-skinned and mild, yet with a distinct flavour and aroma. Like the English cucumber, Lebanese cucumbers are nearly seedless.
- East Asian cucumbers are mild, slender, deep green, and have a bumpy, ridged skin. They can be used for slicing, salads, pickling, etc., and are available year-round. They are usually burpless as well.
- Persian cucumber, which are mini, seedless, and slightly sweet, are available from Canada during the summer, and all year-round in the US. Easy to cut and peel, it is on average 4–7 in. long. They are commonly eaten chopped up in plain yogurt with mint or sliced thin and long with salt and lemon juice. Vines are parthenocarpic, requiring no pollinators for fruit set.
- Beit Alpha cucumbers are small, sweet parthenocarpic cucumbers adapted to the dry climate of the Middle East.
- Apple cucumbers are short, round cucumbers grown in New Zealand and parts of Europe, known for their light yellow-green colour and mildly sweet flavour. When mature, the fruit may grow tiny spines, and contains numerous edible green seeds. The fruit is usually eaten raw, with skin.
- *Schälgurken* are eaten in Germany. Their thick skins are peeled and then they braised or fried, often with minced meat or dill. They are often known by the term 'Schmorgurken'.
- *Dosakai* is a yellow cucumber available in parts of India. These fruits are generally spherical in shape. It is commonly cooked as curry, added in *sambar* or soup, *daal* and also in making *dosa-aavakaaya* (Indian pickle) and chutney; it is also grown and available through farms in Central California.
- *Kekiri* is a smooth skinned cucumber, relatively hard, and not used for salads. It is cooked as spicy curry. It is found in dry zone of Sri Lanka. It becomes orange coloured when the fruit is matured.
- In May 2008, British supermarket chain Sainsbury's unveiled the 'c-thru-cumber', a thin-skinned variety that reportedly does not require peeling.

Armenian

- Armenian cucumbers (also known as yard long cucumbers) are fruits produced by the plant *Cucumis melo* var. *flexuosus*. This is not the same species as the common cucumber (*Cucumis sativus*) although it is closely related. Armenian cucumbers have very long, ribbed fruit with a thin skin that does not require peeling, but are actually an immature melon. This is the variety sold in Middle Eastern markets as "pickled wild cucumber". In North America, the term "wild cucumber" also refers to plants in the genus *Marah*.

Taste

The human olfactory response to cucumbers appear to vary. Most people report a mild, almost watery flavour or a light melon taste, while a small but vocal minority report a highly repugnant taste—some say almost perfume-like. Cucumbers vary in bitterness, even from the same plant. This bitter taste is attributed to the chemical compound Cucurbitacin C. Cucurbitacin is poisonous to livestock, especially sheep.

Management Practices for Avoiding Cucumber Problems

Diseases, pests, and disorders remain a threat to efficient cucumber production and good profits by lowering yields, reducing quality, and making harvests unreliable. The risk of these problems can be reduced by wise management.

- Keep records.
- Select adapted cultivars with resistance to downy mildew, powdery mildew, anthracnose, scab, angular leafspot, and cucumber mosaic.
- Practice a 12-month, noncucurbit crop rotation.
- Select fertile, well drained fields.
- Sample fields annually for pH, nutrients, nematodes, soil insects, and weeds.
- Use certified seed treated with fungicides and insecticides.
- Seed 0.5-1.0-inch deep on raised beds after soils warm to 60° F.
- Place one strong colony of bees per acre after blooms are present.
- Apply a general purpose fungicide shortly after emergence and include an insecticide as needed.
- Diagnose all leafspots and plant abnormalities promptly.
- Use a high pressure (250 + psi) sprayer for fungicide and insecticide applications.
- Irrigate fields to ensure 1.0-inch of water each week.
- Monitor migration of the pickleworm and spray as needed.
- Reapply nitrogen and potassium after leaching rains.

- Pick three times a week.
- Avoid injuries during harvesting and handling.
- Use chlorine in all dump, wash, and handling water.

Poor stands may be caused by cold or wet soils, improper seeding depth or fertilization, residual herbicides, insects, and damping-off caused by several soil inhabiting fungi such as *Pythium spp., Rhizoctonia spp., Fusarium spp.,* and some leafspotting fungi during wet periods. CONTROL: Follow general management practices, replant.

Seedcorn maggot, adult, larva, pupa *(Delia platura)* and injury to seedlings. In the spring adults seek germinating seeds to lay eggs. Maggot injury is associated with cold, wet, organic soils. CONTROL: Apply soil insecticide, replant.

Wireworm larva and adults *(Conoderus vespertinus* and C. *falli)* and injury to small plant. Wireworms are associated with fields previously in small grain or not in row crops. CONTROL: Avoid problem fields and use soil insecticide.

Herbicide injury from residues carried over from previous crop, most often corn. Herbicide injury is fairly common as there are many possibilities for misuse. CONTROL: Keep records of herbicides used in fields; select proper herbicide(s), application method and rates; check equipment calibration; avoid drift; clean equipment properly.

Cold injury is very serious on cucumbers. It is most likely to show up in low areas. CONTROL: Follow general management practices, replant.

Fertilizer injury often occurs with banded fertilizer. Symptoms include poor stands, marginal foliar burning, and browning of roots. CONTROL: Adjust equipment properly, broadcast fertilizer according to soil test reports, reapply fertilizers if needed as determined by soil test or soluble salts.

Serpentine leafminer larva, pupa and adults *(Liriomyza spp.)* and injury. Their presence is common but serious injury is usually confined to small plants. CONTROL: Early detection and prompt insecticide sprays.

Cucumber beetle damage. This insect overwinters as adults and harbors the wilt bacterium. Typically in early spring their feeding causes reduced stands, distorted plants, and wilt, especially on small seedlings. CONTROL: Early detection and prompt insecticide sprays.

Spotted and striped cucumber beetle adults and larva *(Diabrotica undecimpunctata howardi and Acalymma vittatum)* on plant with bacterial wilt. The beetles are overwintering sources and vectors of the bacterial wilt pathogen. Wilt often involves only a portion of a plant. CONTROL: Follow general management practices, early detection of beetles and prompt insecticide sprays. Frequent sprays are required.

Squash bug adults, eggs, and nymphs *(Anasa tristis)* and plant damage. Squash bugs are serious pests of squash and may occur on other cucurbits. Adults and nymphs suck plant juices which blackens and dries out vines. CONTROL: Apply insecticide sprays against the immature stages.

Aphid colonies *(Aphis gossypii* and others) and damage. Winged and wingless aphids occur on the underside of leaves. These sucking insects cause little feeding injury, however they spread viruses that cause mosaic. CONTROL: Apply insecticides if populations are very large. This is not effective for controlling mosaic.

Pickleworm *(Diaphania nitidalis)* and characteristic entrance hole with exudate. The movement of pickleworm moths into North Carolina and their presence in specific fields are monitored with sex pheromone traps or occurrence on squash plants. CONTROL: Begin spraying when migration is detected.

Pickleworm *(Diaphania nitidalis)* and relatives:

a. Early instar pickleworm. Note brown head and black spots.
b. Late instar pickleworm. Note brown head and green body.
c. Melonworm *(Diaphania hyalinata).* Note white stripes on top of body.
d. Pickleworm. Again note brown head with a black notch on the rear of the head capsule and the green body.
e. Squash vine borer *(Melittia cucurbitae).* Note the black head and the white, deep segmented body.
f. Squash vine borer moth. These moths visit curcurbits during the day.

Fertilizer deficiency. Characteristic symptoms of manganese deficiency. CONTROL: Apply fertilizers according to annual soil test reports.

Soybean thrips *(Sericothrips variabilis)* and injury. Feeding winged and wingless, yellow or striped, narrow bodied thrips cause silvery blotches on leaves and reduced plant vigor, especially on small plants during cool, wet periods. CONTROL: Scout fields for presence of thrips and spray when infestations occur. Use a lOx pocket magnifier and white paper to detect thrips.

Two spotted spider mite adult and egg *(Tetranychus urticae)* and foliar injury Mites suck plant juices causing yellowed leaves and poor plant growth. With population buildup during hot, dry periods, fine webbing may be observed. CONTROL: Same as for soybean thrips.

Cabbage looper adult, pupa and larva *(Trichoplusia ni)* and injury. Foliar feeding by loopers can strip leaves and delay growth of small plants in late plantings. CONTROL: Check traps for this migrating insect, scout fields and spray when young larvae are seen.

Deformed fruit often results from poor pollination, extreme weather (too hot, dry, wet, cold) or nutrition levels (too much or little fertilizer). CONTROL: Place one bee colony per acre after flowering, fertilize according to soil test reports, and irrigate.

Honey bee *(Apis mellifera)*. Honey bees are essential for successful cucumber pollination and production. At least one strong colony per acre is needed to ensure adequate bee activity. Colonies should be placed in the field after the onset of significant flowering.

Gummy Stem Blight is common and is caused by the seed-borne fungus *Mycosphaerella citrullina and* the soil-borne fungus *Phoma exiqua*. All cucurbits are susceptible, and depending on the weather, stem and leafspotting may occur on young plants, especially in late plantings. CONTROL: Follow general management practices; start spray schedule at first appearance.

Anthracnose is common and is caused by the seedborne fungus *Colletotrichum lagenarium*. Most cucurbits are susceptible. Most commercial cucumber varieties have some resistance. Depending on weather, leaf and fruit spotting may occur on young plants, especially in late plantings. CONTROL: Follow general management practices; start spray schedule at first appearance.

Angular Leafspot is infrequent and is caused by the seed-borne bacterium *Pseudomonas lachrymans* which can also affect squash and pumpkins. Warm, wet weather may cause stem and leaf spots in midseason. CONTROL: Follow general management practices; start spray schedule at first appearance; do not enter fields when plants are wet.

Powdery Mildew is caused by two air-borne fungi *Erysiphe cichoracearum and Sphaerotheca fulginea,* but is rarely seen since most commercial cucumbers have excellent resistance. Powdery mildew is often serious in dry weather and is the main cause of early dying of cantaloupes, pumpkins, squash, and greenhouse cucumbers. CONTROL: Resistant cultivars; sprays; because of fungicide specificity and possible development of resistance in the fungus, switch fungicides if spraying is not effective.

Scab is caused by the seed-borne fungus *Cladosporium cucumerinum* which also affects muskmelons. It is rare because most commercial cucumbers have excellent resistance. Cool night temperatures and high humidities are required. In the active phase, dark, olive-green, velvety mold covers the spot. Leaves, fruit, and young terminals may be affected. CONTROL: Use resistant varieties, spray at first appearance.

Downy Mildew is caused by the air-borne fungus *Pseudoperonospora cubensis.* Although serious on late season cantaloupes, pumpkins, squash, watermelon, and other cucurbits, it is seldom a problem on cucumbers because most commercial varieties have adequate resistance. The fungus overwinters in Florida, and each year the spores are blown northward and arrive in North Carolina from May to August, if at all. Typically its arrival coincides with summer rains. CONTROL: Resistant cultivars; start spray schedule at first appearance.

Target Spot is caused by the air-borne fungus Co*rynespora cassiicola* which attacks many different plants. It is more common on cucumbers in greenhouses than in fields. CONTROL: Follow general management practices; start spray schedule at first appearance.

Scab on fruit.

Belly Rot is caused by the fungus *Rhizoctonia solani,* present in agricultural soils. It causes damping-off of many crops including cucumbers. Belly rot is common, but its occurrence and intensity are unpredictable. CONTROL: Avoid problem fields; apply fungicide to soil prior to fruit set; keep harvested cucumbers cool, well ventilated; use chlorine in wash.

Bacterial Wilt is caused by the bacterium *Erwinia tracheiphila.* It overwinters in and is spread by cucumber beetles. Often seen in early to mid-season cucumbers and when temperatures are cool, it occurs on muskmelon and to a lesser extent on squash and pumpkin; watermelons may be immune. CONTROL: Control the cucumber beetles, spray.

Root-Knot is caused by *Meloidogyne spp.* These nematodes are usually present in agricultural fields and affect many crops. They are especially damaging on plants growing in light, sandy soils when counts are high. CONTROL: Promptly disk all crops after harvest, treat soil with a fumigant nematicide when warranted by assay report.

Mosaics are caused by viruses. Depicted is Tomato Ringspot, fairly common in the northwestern part of the state. The virus resides in wild brambles and is transmitted by some nematodes. Most commercial cultivars have excellent resistance to Cucumber and Watermelon Mosaic. Until recently mosaics were rarely seen in the eastern counties. Because Zucchini Yellow Mosaic (closely related to Watermelon Mosaic) may be introduced, it is important to identify all mosaics. Mosaic viruses are transmitted by aphids and related insects. CONTROL: Select resistant cultivars; avoid problem fields; start spray schedule with stylet oil and insecticides at the seedling stage; use aluminum plastic mulch.

Cottony Leak is caused by the fungus *Pythium spp.* Present in agricultural soils; it causes damping-off of many crops including cucumbers. Cottony leak is common, but its occurrence and intensity are unpredictable. CONTROL: Avoid problem fields; apply fungicide sprays to soil and foliage prior to fruit set; keep harvested cucumbers cool, well ventilated; use chlorine in wash.

Spring Onions

A scallion (UK: Spring Onion) is one of various *Allium* species, all of which have hollow green leaves (like the common onion), but which lack a fully developed root bulb. It has a relatively mild onion flavour, and is used as a vegetable, either raw or cooked. Many other names are used, including green onion, spring onion, salad onion, table onion, green shallot, onion stick, long onion, baby onion, precious onion, yard onion, gibbon, syboe or scally onion.

The words scallion and shallot are related and can be traced back to the Greek askolonion as described by the Greek writer Theophrastus. This name, in turn, seems to originate from the name of the town of Ashkelon. The plant itself apparently came from farther east of Europe.

Types

The Welsh onion (*Allium fistulosum*) does not form bulbs even when mature, and is grown in the West almost exclusively as a scallion or salad onion, although in Asia this species is of primary importance and used both fresh and in cooking. “Scallion” is also used for young plants of the common onion (*A. cepa* var. *cepa*) and shallot (*A. cepa* var. *aggregatum*, formerly *A. ascalonicum*), harvested before bulbs form, or sometimes when slight bulbing has occurred. Most of the cultivars grown in the West primarily as salad onions or scallions belong to *A. cepa* var. *cepa*. Other species sometimes used as scallions include *A.* ×*proliferum* and *A.* ×*wakegi*.

Species and cultivars which may be called "scallions" include:

- *Allium cepa*
    - o "White Lisbon"
    - o "White Lisbon Winter Hardy" – an extra-hardy variety for overwintering
    - o *Calçot*
- *Allium chinense*
- *Allium fistulosum*
- *A.* ×*proliferum*
- *Allium* ×*wakegi*

Uses

Figure: *Chopped scallions*

Harvested for their taste, they are milder than most onions. They may be cooked or used raw as a part of salads, salsas, or Asian recipes. Diced scallions are used in soup, noodle and seafood dishes, as well as sandwiches, curries or as part of a stir fry. In many Eastern sauces, the bottom half-centimetre (quarter-inch) of scallion roots is commonly removed before use.

In Mexico and the Southwest United States, *cebollitas* are scallions that are sprinkled with salt and grilled whole for cheese and rice . Topped with lime juice, they typically serve as a traditional accompaniment to *asado* dishes. In Catalan cuisine, *calçot* is a variety of green onion traditionally eaten in a *calçotada* (plural: *calçotades*).

A popular gastronomic event of the same name is held between the end of winter and early spring, where *calçots* are grilled, dipped in *salvitxada* or romesco sauce, and consumed in massive quantities.

In Vietnam, Welsh onion is important to prepare *dýa hành* (fermented onions) which is served for T¿t, the Vietnamese New Year. A kind of sauce, *má hành* (Welsh onion fried in oil), is used in dishes such as *cõm t¥m, bánh ít, cà tím nýÛng*, and others. Welsh onion is the main ingredient in the dish *cháo hành*, which is a rice porridge dish to treat the common cold.

In India it is eaten as an appetizer (raw) with main meals. In north India Coriander, Mint and Green Onion Chutney is made using Scallions (raw).

In southern Philippines, it is ground in a mortar along with some ginger and chili pepper to make a native condiment called wet *palapa*, which can be used to spice up dishes, or topped in fried or sun dried food. It could also be used to make the dry version of *palapa*, which is stir fried fresh coconut shavings and wet palapa.

During the Passover meal (Seder), Persian Jews lightly and playfully strike family members with scallions when the Hebrew word *dayenu* is read, symbolizing the whips endured by the Israelites under the ancient Egyptians.

Spring Onion Pests and Diseases

Spring Onions are not particularly affected by pests and diseases as they are quick growing and do not give much time for problems but some can occur and often they are similar to onion. The possible problems are:-

Thrips attack the foliage and eat the leaves from the inside and the plants will eventually die. It is possible to use a detergent spray but the best method is to take out and destroy the plants or a regular basis.

Onion fly are laid in the neck of the onion and will tunnel into the bulb and destroy it, occurs mainly in large onions. The infected onion should be removed and destroyed and the infected area turned over so that predators can feed on them. Use crop rotation to finish the job

Eel worm is a microscopic bug that bores into the onion and makes it swell and go misshapen remove and destroy the onion and do not plant in the same area for several years, the longer the better but at least 3 years, other vegetables like carrots and beans should also not be sown there.

Red Onions

Red onions, sometimes called purple onions, are cultivars of the onion with purplish red skin and white flesh tinged with red.

These onions tend to be medium to large in size and have a mild to sweet flavour. They are often consumed raw, grilled or lightly cooked with other foods, or added as colour to salads. They tend to lose their redness when cooked.

Red onions are available throughout the year. The red colour comes from anthocyanidins such as cyanidin. Red onions are high in flavonoids.

They can be stored 3 to 4 months at room temperature.

Figure: *A red onion*

Red Onion of Turda

The red onion from Turda (Cluj County, Central Romania) (Romanian: "Ceapa de Turda") is a local variety of red onion with light sweeter taste and particular aroma. The area of cultivation encompass the lower Arie° valley and the middle Mure° valley. Turda onion bulbs are traditionally intertwined into long strings (1–2 m) for marketing purposes and can be found at the traditional markets all over central Romania. "Turda Red Onion" is usually served fresh, as a salad or part of mixed salads and especially as a compulsory garnish for the traditional bean-and-smoked ham soups. Red onions are sometimes used as treatments for colds, diarrhea, intestinal parasites, gallstones, and rheumatism.

Red Onion of Tropea

The red onion from Tropea, Italy, (Italian: "Cipolla Rossa di Tropea") is a particular variety of red onion which grows in a small area of Calabria in southern Italy named "Capo Vaticano" near the city of Tropea. This onion has a stronger and sweeter aroma and the inner part is juicier and whiter than other red onions and it is possible to make a marmalade with it. In March 2008, the European Union registered the Protected Designation of Origin mark for the onions produced in this particular area.

Figure: *Cross sections of red onions*

Onion Pests and Diseases

Don't forget to practice something that never goes out of fashion — crop rotation. That's right, be up with the best, rotate your onions with other vegetable crops to prevent a buildup of disease and insect pests. A good suggestion is onions, then carrots the next season, then potatoes, then onions again. These vegetables all have totally different pests and diseases so can't cross infect each other.

Onion Pests

These are few and far between... but there's a couple of these nasty little blighters, or often a busload of these freeloaders you need to watch out for...

Onion Fly: These are not only onion pests, they also attack all members of the allium family, such as leeks, garlic and ornamental alliums.

Onion fly lay eggs which hatch into maggots and eat into onion seedlings usually between the leaves and small bulbs or roots. In older onions the onion fly maggots eat away at the bottom of the onion bulbs.

These onion pests will eventually cause the onion leaves to go pale, wilt then die off, the bulb, or at least the inside of it to become rotten and stinky and the plant to keel over and die. The onion fly looks similar to a small house fly and can be found flying around early summer as it lays its eggs in seedlings or the soil at the base of plants.

After 3 days the eggs hatch and the larvae tunnel into the onion, where they dine non-stop for 3 weeks growing into full-sized 8mm long reddish brown maggots.

Fat and happy they burrow back into the soil and pupate, emerging as a new generation of flies 17 days later.

At the end of summer after a few generations, the final pupae over-winter until next spring.

Lesser Bulb Fly: These prefer narcissus bulbs but occasionally can be an onion pest.

They have a similar lifestyle pattern as the onion fly, but are slightly smaller at about 5mm.

With onion fly and lesser bulb fly, say goodbye to infected onions, but future plants can be protected a couple of ways:

1. Pull up and destroy all plants infected with these onion pests at the time you know they are in the bulbs and before they have burrowed back into the soil. The fact that these onion pests work in patches or bulb by bulb down a row, makes it easier to spot plants in trouble and get them out. Scattering onion planting throughout the garden helps, and removing the soil is also a good idea.
2. Press down the soil around seedlings well, cover with paper, cloth or other insect-proof material to stop the onion fly from laying its eggs into the soil.
3. Spreading a fine layer of sand or wood ash around the onion plants is also a good deterrent to adult onion flies from laying their eggs at the base of the plants.

Thrips: These onion pests can wreck havoc in onion cultivation. They cut the epidermis of leaves or stems and suck the plant sap. This results in white silvery blotches on deformed leaves.

Thrips insects are female and can reproduce without a male. They thrust the eggs into leaves that hatch in five to ten days.

Onion Diseases

Damp and often warm conditions can cause a few types of onion leaf and bulb rot and infection. These onion diseases can occur during cultivation or storage. Here's a list of the most common onion diseases:

Botrytis: There are three botrytis fungus species that affect onions:

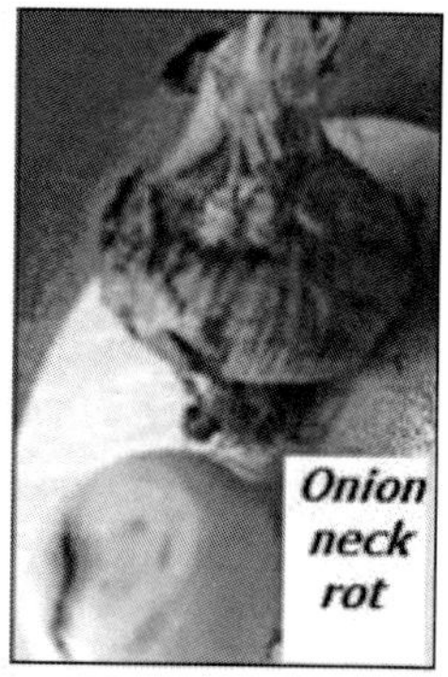

Leaf Blight (Botrytis Squamosa): This is popularly known as blast because it spreads so rapidly particularly in prolonged wet conditions. It causes lesions on the onion leaves which penetrate right through to the inside of the leaves.

Leaf blight looks like white or pale gold specks and eventually it kills the leaves which in turn damages the whole plant, causing it to rot and die.

Onions are most vulnerable to leaf blight when in full leaf growth and as the bulbs are forming. Dig up infected plants and burn, bury deeply away from garden, or seal in bags and remove.

Once leaf blight has a strong hold, it is hard to kill all the spores, but if you suspect infection, spray with an organic fungicide. Also, make sure you buy fresh seed for the next season.

Leaf Fleck: (Botrytis cinerea) is a mild fungus and only causes surface pale green to white flecks on onion leaves. It does not damage the plants and is slow to spread.

Neck Rot: (Botrytis allii) This affects stored onions, although it originates from the soil where the onions grew. The neck of the onion first goes watery, then brown, and then hardens to wizened papery leaves. Often with high humidity there is a grey mould growing between the onion layers.

Onion neck rot only attacks stored onions with tops that have taken too long to dry or have not dried sufficiently before being stored.

Avoiding thick necks on onions is important, so don't encourage growth by fertilising and irrigating late in the season when the bulbs should be maturing and tops drying.

Quickly destroy any stored onions you find with signs of neck rot and always inspect your stored produce regularly.

Downy Mildew: This onion disease is fungal, and the young plants get infected in cool and damp conditions.

A white mold appears and spreads down the entire sheath, eventually causing the plant to fall over.

This can be controlled by treating young plants with an organic fungicide in weekly intervals till bulbing develops.

Onion White Rot: Of all the onion pests and diseases, onion white rot really will make growers "mad-dog mean", as Clint Eastwood would say.

So far impossible to eradicate, this disease is caused by the fungus *Sclerotium cepivorum,* which produces sclerotia — tiny dot sized black fungal bodies that can infect all members of the Allium or Onion family.

Sclerotia can hang around, seemingly dormant, in soil for up to 15 years, so it's important to take serious steps, such as:

- If possible check your source if you are planting onion sets or seedlings, and if you are in any doubt, grow onions from seeds only.
- If you've identified onion white rot, remove the infected plants and burn or seal tightly and dispose of — certainly don't compost.
- Keep any future onions a safe distance, at least 5m (16ft) away from any previous infected spots.
- Watch your shoes — don't walk on previously affected areas and then onto fresh soil because it's easy to transfer the dormant sclerotia.

The first symptoms of onion white rot are the green foliage turning yellow then dying. When the plant is pulled out there are no roots, they have been destroyed and there is nasty white mould in their place. Within that mould lurk the sclerotia.

Pink root: The symptoms of this fungus really are pink roots. Only the roots and sometimes the outer layers of the bulbs get infected. They show up as pale pink first, darkening to deep reddy purple.

Pink root will actually kill the roots, because they shrivel up and go black and die off. This will stunt the hapless onions, but not necessarily kill them totally.

Crop rotation is the best defence against onion pink root.

Purple blotch: In case of Purple Blotch, purple spots develop on the leaves under moist conditions, and in hot (21 to 30 degree Celsius) and humid climates.

This disease is very widespread and can be controlled by lowering planting density, good drainage and routine (weekly) spray of organic fungicides if necessary.

Smudge or Anthracnose: Hot and humid conditions cause the fungus anthracnose, which mostly affects mature and stored onion bulbs. The lower part of the leaves develop sunken pale patches then turn greyish and darker. The leaves become twisted due to this infection. It is important to have healthy growing conditions using compost and good drainage. Destroy any infected plants and rotate area with other resistant crops for at least 2 years. If caught early try an organic fungicide such as baking soda, and see fungal solutions at Organic garden pest control

Smut: This fungus can attack seedlings just as the germinate, but doesn't always kill them. Onions that survive are not so healthy and often succumb to some other disease later, even when harvested and stored.

Onion smut starts with dark lesions on the leaves of young seedlings — healthy older plants are not affected by smut spores.

The smut affected seedlings grow on to produce onions with thickened, blistered plant walls. The onions are stunted and bulbs are mishappened sometimes with quite dark blister patches. Follow the advice for smudge/anthracnose above for controlling onion smut.

Stemphylium blight: In Stemphylium leaf blight, the symptoms, causes and remedy are similar to Purple Blotch with the difference that the blight lesions are light yellow to brown.

Mushrooms

A mushroom (or toadstool) is the fleshy, spore-bearing fruiting body of a fungus, typically produced above ground on soil or on its food source. The standard for the name "mushroom" is the cultivated white button mushroom, *Agaricus bisporus*; hence the word "mushroom" is most often applied to those fungi (Basidiomycota, Agaricomycetes) that have a stem (stipe), a cap (pileus), and gills (lamellae, sing. lamella) or pores on the underside of the cap. These pores or gills produce microscopic spores that help the fungus spread across the ground or its occupant surface.

"Mushroom" describes a variety of gilled fungi, with or without stems, and the term is used even more generally, to describe both the fleshy fruiting bodies of some Ascomycota and the woody or leathery fruiting bodies of some Basidiomycota, depending upon the context of the word.

Forms deviating from the standard morphology usually have more specific names, such as "puffball", "stinkhorn", and "morel", and gilled mushrooms themselves are often called "agarics" in reference to their similarity to *Agaricus* or their place Agaricales. By extension, the term "mushroom" can also designate the entire fungus when in culture; the thallus (called a mycelium) of species forming the fruiting bodies called mushrooms; or the species itself.

Edible Mushroom

Figure: *White mushrooms ready for cooking. While common, they are just one of the many types of mushrooms cultivated and eaten.*

Figure: *Picked edible mushrooms in a basket*

Edible mushrooms are the fleshy and edible fruit bodies of several species of macrofungi (fungi which bear fruiting structures that are large enough to be seen with the naked eye). They can appear either below ground (hypogeous) or above ground (epigeous) where they may be picked by hand. Edibility may be defined by criteria that include absence of poisonous effects on humans and desirable taste and aroma.

Edible mushrooms are consumed by humans as comestibles for their nutritional value and they are occasionally consumed for their supposed medicinal value. Mushrooms consumed by those practicing folk medicine are known as medicinal mushrooms. While hallucinogenic mushrooms (e.g. Psilocybin mushrooms) are occasionally consumed for recreational or religious purposes, they can produce severe nausea and disorientation, and are therefore not commonly considered edible mushrooms.

Edible mushrooms include many fungal species that are either harvested wild or cultivated. Easily cultivatable and common wild mushrooms are often available in markets, and those that are more difficult to obtain (such as the prized truffle and matsutake) may be collected on a smaller scale by private gatherers. Some preparations may render certain poisonous mushrooms fit for consumption.

Before assuming that any wild mushroom is edible, it should be identified. Proper identification of a species is the only safe way to ensure edibility. Some mushrooms that are edible for most people can cause allergic reactions in some individuals, and old or improperly stored specimens can cause food poisoning. Deadly poisonous mushrooms that are frequently confused with edible mushrooms and responsible for many fatal poisonings include several species of the *Amanita* genus, in particular, *Amanita phalloides*, the *death cap*. Mushrooms growing in polluted locations can accumulate pollutants such as heavy metals

Commercially Cultivated

Figure: *Commercial cultivated Japanese edible mushroom species. Clockwise from left, enokitake, buna-shimeji, bunapi-shimeji, king oyster mushroom and shiitake.*

Mushroom cultivation has a long history, with over twenty species commercially cultivated. Mushrooms are cultivated in at least 60 countries with China, the United States, Netherlands, France and Poland being the top five producers in 2000.

Commercially Harvested Wild Edibles

Some species are difficult to cultivate; others (particularly mycorrhizal species) have not yet been successfully cultivated. Some of these species are harvested from the wild, and can be found in markets. When in season they can be purchased fresh, and many species are sold dried as well.

Fungiculture

Fungiculture is the process of producing food, medicine, and other products by the cultivation of mushrooms and other fungi. The word is also commonly used to refer to the practice of cultivating fungi by leafcutter ants, termites, ambrosia beetles, and marsh periwinkles.

Mushrooms are not plants, and require different conditions for optimal growth. Plants develop through photosynthesis, a process that converts atmospheric carbon dioxide into carbohydrates, especially cellulose. While sunlight provides an energy source for plants, mushrooms derive all of their energy and growth materials from their growth medium, through biochemical decomposition processes. This does not mean that light is an irrelevant requirement, since some fungi use light as a signal for fruiting. However, all the materials for growth must already be present in the growth medium. Mushrooms grow well at relative humidity levels of around 95-100%, and substrate moisture levels of 50 to 75%.

Instead of seeds, mushrooms reproduce asexually through spores. Spores can be contaminated with airborne microorganisms, which will interfere with mushroom growth and prevent a healthy crop.

Mycelium, or actively growing mushroom culture, is placed on a substrate—usually sterilized grains such as rye or millet—and induced to grow into those grains. This is called inoculation. Inoculated grains are referred to as spawn. Spores are another inoculation option, but are less developed than established mycelium. Since they are also contaminated easily, they are only manipulated in laboratory conditions with a laminar flow cabinet.

Techniques

All mushroom growing techniques require the correct combination of humidity, temperature, substrate (growth medium) and inoculum (spawn or starter culture). Wild harvests, outdoor log inoculation and indoor trays all provide these elements.

Wild Harvesting

Due to its climate, the Pacific Northwest of the USA produces commercially valuable mushrooms. Valued species include:

- American matsutake or pine mushroom (*Tricholoma magnivelare*)
- Chanterelles (*Cantharellus formosus*, *Cantharellus subalbidus*, and *Cantharellus cibarius*)
- Horn of plenty (*Craterellus cornucopioides*)
- Boletes (*Boletus edulis* and others)
- Truffles (*Tuber gibbosum* and *Leucangium carthusiana*)
- Hedgehogs / "spreading-hedgehog mushroom" (*Hydnum repandum*)
- Edible morel (*Morchella esculenta*)
- Coral tooth mushroom (*Hericium abietis*)
- Shaggy parasol (*Lepiota rhacodes*)
- Black picoa (*Picoa carthusiana*)
- Cauliflower mushroom (*Sparassis crispa*)

Mushroom gatherers have few requirements to begin business. Gatherers only need to supply funds for possible park fees, knowledge for identifying mushrooms and gathering time.

There are significant disadvantages to relying on natural mushroom production. These sales may be unregulated, placing buyers at risk for buying toxic or inedible mushrooms. By honest error, harvests may include toxic or inedible species. No controls exist to regulate the quality or frequency of harvests, since gatherers rely on favourable natural conditions and weather to produce fruiting. Conflicts may arise between competing gatherers trying to harvest from the same location.

State parks in the Pacific Northwest or elsewhere may charge fees for mushroom gathering permits. Appalachia also produces edible wild mushrooms, including chanterelles and morels. Pickers may sell directly to distributors, restaurants, or sell their harvest through roadside stands wherever a natural supply of mushrooms is plentiful.

While there may be concern that harvesting wild mushrooms may exploit or damage a natural environment, harvesting wild mushrooms is different from harvesting wild plants, fishing or hunting animals. In these last three cases, removing individuals decreases the ability of a wild population to reproduce, since fewer adults remain. Removing adults leaves fewer individuals capable of reproducing and reduces genetic diversity. Harvesting wild mushrooms removes only fruiting bodies and their attached spores. However, the fruiting bodies (mushrooms) have likely dropped spores before harvest time, or will likely drop them en route to the harvester's destination, further expanding the fungi's habitat. Arguably, the practice of mushroom harvesting may actually help the species being harvested. While truffles also represent the fruiting body of a larger underground network, they are an exception, since they rely on animal spore dispersion. Additionally, reproduction and propagation can still occur by propagation of the parent mycelium. Harvesting removes none of the parent mycelium, which remains intact underground.

Outdoor Logs

Mushrooms can be grown on logs placed outdoors in stacks or piles, as has been done for hundreds of years. Sterilization is not performed in this method. Since production may be unpredictable and seasonal, less than 5% of commercially sold mushrooms are produced this way. Here, tree logs are inoculated with spawn, then allowed to grow as they would in wild conditions. Fruiting, or pinning, is triggered by seasonal changes, or by briefly soaking the logs in cool water. Shiitake and oyster mushrooms have traditionally been produced using the outdoor log technique, although controlled techniques such as indoor tray growing or artificial logs made of compressed substrate have been substituted.

Shiitake mushrooms grown under a forested canopy are considered non-timber forest products In the Northeast shiitake mushrooms can be cultivated on a variety of hardwood logs including oak, American beech, sugar maple and hophornbeam. Softwood should not be used to cultivate shiitake mushrooms. The resin of softwoods will often times inhibit the growth of the shiitake mushroom making it impractical as a growing substrate. In order to produce shiitake mushrooms, 3 foot hardwood logs with a diameter ranging between 4 and 6 inches are inoculated with the mycelium of the shiitake fungus. Inoculation is completed by drilling holes in hardwood logs, filling the holes with cultured shiitake mycelium or inoculum, and then sealing the filled holes with hot wax. After inoculation, the logs are placed under the closed canopy of a coniferous stand and are left to incubate for 12 to 15 months. Once incubation is complete, the logs are soaked in water for 24 hours. 7 to 10 days after soaking, shiitake mushrooms will begin to fruit and can be harvested once fully ripe.

Indoor Trays

Indoor growing provides the ability to tightly regulate light, temperature and humidity while excluding contaminants and pests.

This allows consistent production, regulated by spawning cycles. This is typically accomplished in windowless, purpose-built buildings, for large scale commercial production.Indoor tray growing is the most common commercial technique, followed by containerized growing. The tray technique provides the advantages of scalability and easier harvesting. Unlike wild harvests, indoor techniques provide tight control over growing substrate composition and growing conditions. Indoor harvests are much more predictable.

According to Daniel Royse and Robert Beelman, "[Indoor] Mushroom farming consists of six steps, and although the divisions are somewhat arbitrary, these steps identify what is needed to form a production system. The six steps are phase I composting, phase II fertilizing, spawning, casing, pinning, and cropping."

Six Phases of Mushroom Cultivation

Phase	*Time span*	*Temperature*	*Key points*
1. Phase I composting	6–14 days		Regulate water and NH_3 content through microbial action. Add fertilizer / additives
2. Phase II composting or pasteurization	7–18 days via composting method, ~2 hours for pasteurization (heat sterilization)		Reduce number of potentially harmful microbes through further composting, or apply heat sterilization. Remove unwanted NH_3.
3. Spawning and growth	14–21 days	75 °F; to 80 °F; must be above 74 °F; for rapid growth. Must be below 80 °F; to 85 °F to avoid damaging mycelia	Add starter culture. Allow mycelium to grow through substrate and form a colony. Depends on substrate dimensions and composition. Finished when mycelium has propagated through entire substrate layer
4. Casing	13–20 days		Promote the formation of promordia, or mushroom pins. Add a top covering or dressing to the colonized substrate. Fertilizing with nitrogen increases yields. Induces pinning
5. Pinning	18–21 days		Earliest formation of recognizable mushrooms from mycelium. Adjusting temperature, humidity and CO_2 will also affect the number of pins, and mushroom size
6. Cropping	Repeated over 7-10 day cycles		Harvest

Complete sterilization is not always required or performed during composting. In some cases, a pasteurization step is not included to allow some beneficial microorganisms to remain in the growth substrate.

Specific time spans and temperatures required during stages 3–6 will vary respective to species and variety. Substrate composition and the geometry of growth substrate will also affect the ideal times and temperatures.

Pinning is the trickiest part for a mushroom grower, since a combination of carbon dioxide (CO_2) concentration, temperature, light, and humidity triggers mushrooms towards fruiting. Up until the point when rhizomorphs or mushroom "pins" appear, the mycelium is an amorphous mass spread throughout the growth substrate, unrecognisable as a mushroom.

Carbon dioxide concentration becomes elevated during the vegetative growth phase, when mycelium is sealed in a gas-resistant plastic barrier or bag which traps gases produced by the growing mycelium. To induce pinning, this barrier is opened or ruptured. CO_2 concentration then decreases from about 0.08% to 0.04%, the ambient atmospheric level.

Substrates

Mushroom production converts the raw natural ingredients into mushroom tissue, most notably the carbohydrate chitin. An ideal substrate will contain enough nitrogen and carbohydrate for rapid mushroom growth. Common bulk substrates include:

- Wood chips or sawdust
- Mulched straw (usually wheat, but also rice and other straws)
- Strawbedded horse or poultry manure
- Corncobs
- Waste or recycled paper
- coffee pulp or grounds
- Nut and seed hulls
- Cottonseed hulls
- Cocoa bean hulls
- Cottonseed meal
- Soybean meal
- Brewer's grain
- Ammonium nitrate
- Urea

Mushrooms metabolize complex carbohydrates in their substrate into glucose, which is then transported through the mycelium as needed for growth and energy. While it is used as a main energy source, its concentration in the growth medium should not exceed 2%. For ideal fruiting, closer to 1% is ideal.

Pests and Diseases

Parasitic insects, bacteria and other fungi all pose risks to indoor production. The sciarid fly or phorid fly may lay eggs in the growth medium, which hatch into maggots and damage developing mushrooms during all growth stages. Bacterial blotch caused by Pseudomonas bacteria or patches of Trichoderma green mold also pose a risks during the fruiting stage. Pesticides and sanitizing agents are available to use against these infestations. Biological controls for insect sciarid and phorid flies have also been proposed.

A recent epidemic of Trichoderma green mold has significantly affected mushroom production: "From 1994-96, crop losses in Pennsylvania ranged from 30 to 100%".

Commercially Cultivated Fungi

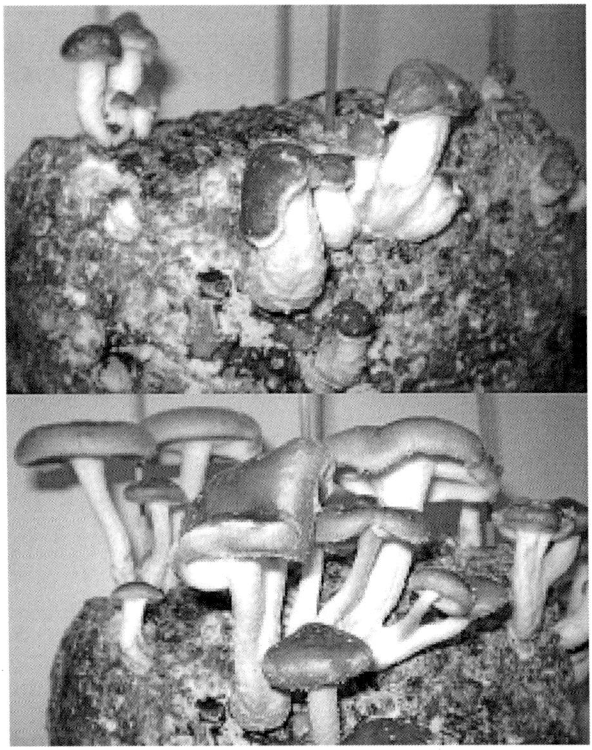

Figure: *Home cultivated* shiitake *developing over approximately 24 hours.*

***Figure:** 9-day old Pleurotus ostreatus mycelium. The structure is a network of millions of interweived threads called hyphae. One can also discern individual spores.*

- *Agaricus bisporus*, also known as champignon and the button mushroom. This species also includes the portobello and crimini mushrooms.
- *Auricularia polytricha* or *Auricularia auricula-judae* (Tree ear fungus), two closely related species of jelly fungi that are commonly used in Chinese cuisine.
- *Flammulina velutipes*, the "winter mushroom", also known as *enokitake* in Japan
- *Hypsizygus tessulatus* (also *Hypsizygus marmoreus*), called *shimeji* in Japanese, it is a common variety of mushroom available in most markets in Japan. Known as "Beech mushroom" in Europe.
- *Lentinus edodes*, also known as shiitake, oak mushroom. *Lentinus edodes* is largely produced in Japan, China and South Korea. *Lentinus edodes* accounts for 10% of world production of cultivated mushrooms. Common in Japan, China, Australia and North America.
- *Pleurotus* species are the second most important mushrooms in production in the world, accounting for 25% of total world production. *Pleurotus* mushrooms are cultivated worldwide; China is the major producer. Several species can be grown on carbonaceous matter such as straw or newspaper. In the wild they are usually found growing on wood.
    - o *Pleurotus cornucopiae*
    - o *Pleurotus eryngii* (king trumpet mushroom)
    - o *Pleurotus ostreatus* (oyster mushroom)

- *Rhizopus oligosporus* - the fungal starter culture used in the production of tempeh. In tempeh the mycelia of *R. oligosporus* are consumed.
- *Sparassis crispa* - recent developments have led to this being cultivated in California.
- *Tremella fuciformis* (Snow fungus), another type of jelly fungus that is commonly used in Chinese cuisine.
- *Tuber* species, (the truffle), Truffles belong to the ascomycete grouping of fungi. The truffle fruitbodies develop underground in mycorrhizal association with certain trees e.g. oak, poplar, beech, and hazel. Being difficult to find, trained pigs or dogs are often used to sniff them out for easy harvesting.
    - *Tuber aestivum* (Summer or St. Jean truffle)
    - *Tuber magnatum* (Piemont white truffle)
    - *Tuber melanosporum* (Périgord truffle)
    - *T.melanosporum* x *T.magnatum* (Khanaqa truffle)
    - *Terfezia sp.* (Desert truffle)
- *Ustilago maydis* (Corn smut), a fungal pathogen of the maize plants. Also called the Mexican truffle, although not a true truffle.
- *Volvariella volvacea* (the "Paddy straw mushroom.") *Volvariella* mushrooms account for 16% of total production of cultivated mushrooms in the world.

Integrated Pest Management in Mushroom Production

Current constraints associated with the use of chemical insecticides and increasing consumer demand for reduced pesticide inputs necessitate the development of novel insect pest control programmes in mushroom production. A new approach to insect pest control in mushroom production is currently being investigated by the division, which, while not eliminating insecticide application, will maximise its effects in combination with cultural practices. The protected, controlled environment associated with mushroom cultivation provides suitable conditions for rapid increases in pest populations. However, the controlled environment also provides the potential to integrate pest control measures with other cultural practices to provide more effective and sustainable insect pest control.

It is possible for growers to achieve optimal insect pest control with rational insecticide use, provided they integrate cultural practices, environmental factors and biological controls. A full understanding of the biology of the pests and the nature of the damage caused is essential for effective integrated pest management.

Presently, sciarids (Lycoriella spp.) are considered the most significant insect pest during mushroom cultivation and adults may be found on mushroom farms throughout the year but more commonly between May - November. Adult flies are small, grey delicate insects with a tendency to run across surfaces rather than fly. Within insulated polythene tunnels they may be observed congregating or "roosting" around the structural "hoops". Efficient pasteurisation destroys immature stages in phase II compost but females may subsequently be attracted to the compost during cooling. Egg laying, can occur before and, up to a week, after spawning.

A second generation of flies emerges from the compost within two to three weeks, which coincides with casing operations. The casing layer provides the medium for the development of subsequent generations that damage the emerging crop.

Mushroom compost is also susceptible to attack by the phorid fly (Megaselia halterata). Adult phorids are brown-black, robust, hump-backed flies with characteristic jerky movements and they are commonly referred to as "scuttle flies". Phorid flies are active during May to November with a peak in September.

Females lay eggs in compost, which has been spawn-running for approximately 7-12 days. Egg laying, continues in the casing layer throughout the remainder of the cropping cycle. The recent withdrawal of the organophosphorus compound diazinon has compounded the difficulties associated with control of this pest and may increase its economic significance.

From the life cycles of the pests outlined above it may be assumed that the potential for infestation is primarily at the site of crop production. However, composters have a responsibility to minimise the period between compost manufacture and delivery, to reduce the possibility of infestation with by invertebrate pests.

Preventative Control Measures

- Growing tunnels should always be properly sealed to prevent adult migration into the house.
- Cracks and joints should be sealed with proprietary sealants.
- Mesh screens or filters should be inserted over fan intakes and vents.

- Personnel and loading doors should also be sealed and entry to production houses, particularly before cropping, should be restricted.
- Casing procedures should take place within the production house and personnel and loading doors should remain closed.
- Post-cropping disinfection of production houses will minimise carry-over of pests from one crop to another. Before spent compost is removed, a fumigant treatment with formaldehyde may be applied.
- Thorough hygiene procedures will eliminate insect breeding and roosting sites. Organic litter should be regularly removed from the production site. Stagnant water pools should not be created following hosing down.
- Fumigation of production houses will be most effective following procedures, which increase crop exposure to flies (i.e. beginning of spawn and case-runs).

Chemical and Biological Control Measures

Experimental work to examine chemical and biological control of sciarids demonstrated that normal application diflubenzuron (Dimilin) at casing reduced sciarid emergence during case-run by 88% and the controlling effects persisted into the period of third flush. This level of control was achieved without the persistent phytotoxic effects on mushroom yield previously noted with diflubenzuron used in conjunction with diazinon. Reduction in mushroom yield at first flush associated with application of diflubenzuron was compensated by increased mushroom yields in subsequent flushes.

Methoprene applied at casing reduced sciarid emergence during case-run by 27% and controlling effects increased to 43% at first flush. However, no controlling effects on sciarid emergence were observed during the third flush. Application of the biological control agent Steinernema feltiae (Nemasys M) reduced the number of immature sciarids during case-run by 89% and at first flush by 94%. However, the level of control decreased to 68% at third flush These biological control agents were most effective when applied following casing stage rather than at early spawn-run.

Introduction of the predatory mite Hypoaspis miles (Hyposure) at case-run reduced the number of immature sciarids during this period by 18%, increasing to 27% at first flush. However, no controlling effects of this biological control agent were observed at third flush.

Avocado

The avocado (*Persea americana*) is a tree native to Mexico and Central America, classified in the flowering plant family Lauraceae along with cinnamon, camphor and bay laurel. Avocado or alligator pear also refers to the fruit, botanically a large berry that contains a single seed.

Avocados are commercially valuable and are cultivated in tropical and Mediterranean climates throughout the world. They have a green-skinned, fleshy body that may be pear-shaped, egg-shaped, or spherical. Commercially, they ripen after harvesting. Trees are partially self-pollinating and often are propagated through grafting to maintain a predictable quality and quantity of the fruit.

Cultivation

Figure: *Persea americana, young avocado plant (seedling), complete with parted pit and roots*

Food and Agriculture		
Country	***Quantity (t) 2011***	***World Rank***
Mexico	1,264,141	1
Chile	368,568	2
Dominican Republic	295,080	3
Indonesia	275,953	4

Contd...

Country	Quantity (t) 2011	World Rank
United States of America	238,544	5
Colombia	215,095	6
Peru	212,857	7
Kenya	201,478	8
Brazil	160,376	9
Rwanda	143,281	10
China	108,500	11

The tree grows to 20 m (66 ft), with alternately arranged leaves 12 centimetres (4.7 in) – 25 centimetres (9.8 in) long. The flowers are inconspicuous, greenish-yellow, 5 millimetres (0.2 in) – 10 millimetres (0.4 in) wide. The pear-shaped fruit is 7 centimetres (2.8 in) – 20 centimetres (7.9 in) long, weighs between 100 grams (3.5 oz) – 1,000 grams (35 oz), and has a large central seed, 5 centimetres (2.0 in) – 6.4 centimetres (2.5 in) long.

The subtropical species needs a climate without frost and with little wind. High winds reduce the humidity, dehydrate the flowers, and affect pollination. When even a mild frost occurs, premature fruit drop may occur, although the Hass cultivar can tolerate temperatures down to "1 °C. The trees also need well-aerated soils, ideally more than 1 m deep. Yield is reduced when the irrigation water is highly saline. These soil and climate conditions are available only in a few areas of the world, particularly in southern Spain, Portugal, Morocco, Crete, the Levant, South Africa, Colombia, Peru, parts of central and northern Chile, Vietnam, Indonesia, parts of southern India, Sri Lanka, Australia, New Zealand, the Philippines, Malaysia, Central America, the Caribbean, Mexico, California, Arizona, Puerto Rico, Texas, Florida, Hawaii, Ecuador and Rwanda. Each region has different types of cultivars.

Harvest and Postharvest

Commercial orchards produce an average of seven tonnes per hectare each year, with some orchards achieving 20 tonnes per hectare. Biennial bearing can be a problem, with heavy crops in one year being followed by poor yields the next. The avocado tree does not tolerate freezing temperatures, and can be grown only in subtropical or tropical climates. There are several cold-hardy varieties planted in the region of Gainesville, Florida, which survive temperatures as low as "6.5 °C (20 °F) with only minor leaf damage.

The avocado is a climacteric fruit (the banana is another), which means it matures on the tree, but ripens off the tree. Avocados used in commerce are picked hard and green and kept in coolers at 3.3 to 5.6 °C (37.9 to 42.1 °F) until they reach their final destination. Avocados must be mature to ripen properly. Avocados that fall off the tree ripen on the ground. Generally, the fruit is picked once it reaches maturity; Mexican growers pick Hass-variety avocados when they have more than 23% dry matter, and other producing countries have similar standards. Once picked, avocados ripen in one to two weeks (depending on the cultivar) at room temperature (faster if stored with other fruits such as apples or bananas, because of the influence of ethylene gas). Some supermarkets sell pre-ripened avocados which have been treated with synthetic ethylene to hasten ripening. The use of an ethylene gas "ripening room", which is now an "industry standard," was pioneered in the 1980s by farmer Gil Henry of Escondido, California, in response to footage from a hidden supermarket camera which showed shoppers repeatedly squeezing hard, unripe avocados, putting them "back in the bin," and moving on without making a purchase. In some cases avocados can be left on the tree for several months, which is an advantage to commercial growers who seek the greatest return for their crop; but if the fruit remains unpicked for too long it falls to the ground.

Breeding

Figure: *A seedless avocado, or cuke, growing next to two regular avocados*

The species is only partially able to self-pollinate because of dichogamy in its flowering. This limitation, added to the long juvenile period, makes the species difficult to breed. Most cultivars are propagated via grafting, having originated from random seedling plants or minor mutations derived from cultivars. Modern breeding programs tend to use isolation plots where the chances of cross-pollination are reduced. That is the case for programs at the University of California, Riverside, as well as the Volcani Centre and the Instituto de Investigaciones Agropecuarias in Chile.

The avocado is unusual in that the timing of the male and female flower phases differs among cultivars. There are two flowering types, "A" and "B". "A" cultivar flowers open as female on the morning of the first day and close in late morning or early afternoon. Then they open as male in the afternoon of the second day. "B" varieties open as female on the afternoon of the first day, close in late afternoon and reopen as male the following morning.

- "A" cultivars: Hass, Gwen, Lamb Hass, Pinkerton, Reed.
- "B" cultivars: Fuerte, Sharwil, Zutano, Bacon, Ettinger, Sir Prize, Walter Hole.

Certain cultivars, such as the Hass, have a tendency to bear well only in alternate years. After a season with a low yield, due to factors such as cold (which the avocado does not tolerate well), the trees tend to produce abundantly the next season. In addition, due to environmental circumstances during some years, seedless avocados may appear on the trees. Known in the avocado industry as "cukes", they are usually discarded commercially due to their small size.

Propagation and Rootstocks

Figure: *Avocado is usually treated with a special technique to assist its sprouting process*

Figure: *A young avocado sprout*

Avocados can be propagated by seed, taking roughly four to six years to bear fruit, although in some cases seedlings can take 10 years to come into bearing. The offspring is unlikely to be identical to the parent cultivar in fruit quality. Prime quality varieties are therefore propagated by grafting to rootstocks that are propagated by seed (seedling rootstocks) or by layering (clonal rootstocks). After about a year of growing in a greenhouse, the young rootstocks are ready to be grafted. Terminal and lateral grafting is normally used. The scion cultivar grows for another 6–12 months before the tree is ready to be sold. Clonal rootstocks are selected for tolerance of specific soil and disease conditions, such as poor soil aeration or resistance to the soil-borne disease (root rot) caused by *Phytophthora*.

Growing Indoors

Usually, avocados are grown from pits indoors. This is often done by removing the pit from a ripe, unrefrigerated avocado. The pit is then stabbed with three or four toothpicks, about one third of the way up. The pit is placed in a jar or vase containing tepid water. It should split in four to six weeks and yield roots and a sprout. If there is no change by this time, the avocado pit is discarded. Once the stem has grown a few inches, it is placed in a pot with soil. It should be watered every few days. Avocados have been known to grow large, so owners must be ready to repot the plant several times.

Diseases

Avocado trees are vulnerable to bacterial, viral, fungal and nutritional diseases (excesses and deficiencies of key minerals). Disease can affect all parts of the plant, causing spotting, rotting, cankers, pitting and discolouration.

Figure: P. americana, *avocado plant flowers*

Hass: While dozens of cultivars are grown, the Hass avocado is today the most common. It produces fruit year-round and accounts for 80% of cultivated avocados in the world.

Gwen: A seedling bred from Hass x Thille in 1982, Gwen is higher yielding and more dwarfing than Hass in California. The fruit has an oval shape, slightly smaller than Hass (100–200 g or 3.5–7.1 oz), with a rich, nutty flavour. The skin texture is more finely pebbled than Hass, and is dull green when ripe. It is frost-hardy down to "1 °C (30 °F).

Lula: A seedling reportedly grown from a "Taft" avocado planted in Miami, Florida, on the property of George Cellon, named after Cellon's wife, Lula. It was likely a cross between Mexican and Guatemalan types. Lula was recognised for its flavour and high oil content and propagated commercially in Florida. It is also very commonly used as a rootstock for nursery production. Hardy to "4 °C (25 °F)

Pinkerton: First grown on the Pinkerton Ranch in Saticoy, California, in the early 1970s, Pinkerton is a seedling of Hass's Rincon. The large fruit has a small seed, and its green skin deepens in colour as it ripens. The thick flesh has a smooth, creamy texture, pale green colour, good flavour and high oil content. It shows some cold tolerance, to "1 °C (30 °F) and bears consistently heavy crops. A hybrid Guatemalan type, it has excellent peeling characteristics.

Reed: Developed from a chance seedling found in 1948 by James S. Reed in California, Reed has large, round, green fruit with a smooth texture and dark, thick, glossy skin. Smooth and delicate, the flesh has a slightly nutty flavour. The skin ripens green. A Guatemalan type, it is hardy to "1 °C (30 °F). Tree size is about 5 by 4 metres (16.4 by 13.1 ft).

B Cultivars

Bacon: Developed by a farmer, James Bacon, in 1954, Bacon has medium-sized fruit with smooth, green skin with yellow-green, light tasting flesh. When ripe, the skin remains green, but darkens slightly, and fruit yields to gentle pressure. It is cold-hardy down to "5 °C (23 °F).

Brogden: Possibly a cross between Mexican and West Indian types, Brogden originated as a seedling grown in Winter Haven, Florida on the property of Tom W. Brogden. The variety was recognised for its cold-hardiness to "5 °C (23 °F) and became commercially propagated as nursery-stock for home growing. It is noted for its dark purple skin at maturity.

Ettinger: A Mexican/Guatemalan cross seedling of Fuerte, this cultivar originated in Israel, and was put into production there in 1947. Mature trees tolerate four hours at "6 °C (21 °F). The fruit has a smooth, thin, green skin that does not peel easily. The flesh is very pale green.

Fuerte: A Mexican/Guatemalan cross originating in Puebla, the Fuerte earned its name, which means strong in Spanish, after it withstood a severe frost in California in 1913. Hardy to "3 °C (27 °F), it has medium-sized, pear-shaped fruit with a green, leathery, easy to peel skin. The creamy flesh of mild and rich flavour has 18% oil. The skin ripens green. Tree size is 6 by 4 metres (19.7 by 13.1 ft).

Monroe: A Guatemalan/West Indian cross that originated from a seedling grown in Homestead, Florida on the property of J.J.L. Phillips, it was patented in 1937 and became a major commercial cultivar due to its cold hardiness and production qualities. The fruit is large, averaging over two pounds (0.91 kg) in weight, has an elliptical shape, and green, glossy skin. Hardy to "3 °C (27 °F).

Sharwil: Predominantly Guatemalan, with some Mexican race genes, Sharwil was selected in 1951 by Sir Frank Sharpe at Redland Bay, southern Queensland, Australia. The name "Sharwil" is an amalgamation of Sharp and Wilson (J.C. Wilson being the first propagator). Scions were sent from Australia to Hawaii in 1966. A medium-sized fruit with rough green skin, it closely resembles the Fuerte, but is slightly more oval in shape. The fruit has greenish-yellow flesh with a rich, nutty flavour and high oil content (20–24%), and a small seed. The skin is green when ripe. It represents more than 57%

of the commercial farming in Hawaii, and represents up to 20% of all avocados grown in New South Wales, Australia. It is a regular and moderate bearer with excellent quality fruit, but is sensitive to frost. Disease and pest resistance are superior to Fuerte.

Zutano: Originated by R.L. Ruitt in Fallbrook in 1926, this Mexican variety is hardy to "4 °C (25 °F). The large, pear-shaped fruit has a shiny, thin, yellow-green skin that peels moderately easily. The flesh is pale green with fibres and has a light flavour.

Other Cultivars

Other avocado cultivars include Spinks. The fruit of the cultivar Florida, grown mostly outside California, is larger and rounder, with a smooth, medium-green skin, and a less-fatty, firmer and fibrous flesh. These are occasionally marketed as low-calorie avocados.

High Priority Exotic Pests

The following are some key exotic pests of avocados, as identified through the development of the Industry Biosecurity Plan for the Avocado Industry. *Additional information is included in the fact sheets.*

The climate of Australian avocado production regions would allow each of these pests to survive, spread and establish, should they be introduced. Any of these pests would have serious consequences should they enter and become established in Australia. Implementing biosecurity measures to control endemic pests will go a long way towards preventing exotic pests from entering and becoming established on your farm.

Avocado Scab

- Sphaceloma perseae.
- Fungal disease causing fruit drop and marking of mature fruit
- Discrete spots appear on the leaves along the midrib, which may merge into star-like patterns
- As disease progresses leaves become distorted and stunted
- Symptoms on fruit appear as corky, raised brown to purplish brown spots which enlarge to form large rough areas
- Long distance dispersal likely through infested propagation material

Avocado Seed Moth

- Stenoma catenifer.
- Pest of avocados in Latin America causing large yield losses

- Early instar larvae are white but become violet in later instars
- Adult moths have a 'C' shape of black dots across the forewings
- Moths form galleries in branches causing them to wither
- Larvae penetrate fruit and destroy seed and pulp through feeding
- Frass can be seen in fruit when opened, and oozing of a white exudate may occur externally

Avocado Sunblotch

- Avocado sunblotch viroid.
- Found sporadically in some areas of Australia
- Fruit are small and deformed with sunken pale blotches
- Discolouration and streaking occurs on stems
- A range of symptoms are seen on leaves though some infected trees remain symptomless
- Spread through infected seed or using infected scions for grafting

Avocado Thrips

- Scirtothrips perseae.
- Emerging insect pest causing large economic losses in California
- Adults are straw coloured with dark lines across the abdomen
- Causes bronzing along leaf veins or over entire leaves under severe infestation
- Fruit are scarred with a characteristic 'alligator skin' appearance
- Scars elongate as fruit develops causing lower quality fruit
- Disseminated over long distances on infested avocado plant material

Laurel Wilt

- Raffaelea lauricola.
- Emerging fungal disease of avocado trees in Florida
- Spread by the tiny (2 mm long) Redbay ambrosia beetle (exotic)
- Leaves wilt and foliage drop may follow
- Xylem tissue becomes dark brown or black
- Dying trees attract beetles in large numbers and small strings of wood fibre may be seen protruding from bore holes

Persea Mite

- Oligonychus perseae.
- Adult females are yellowish green with several dark spots in the abdominal region
- Males are smaller with yellow, pear-shaped bodies
- Nests form along midribs and veins on the underside of leaves
- Symptoms can also be seen as necrotic spots covered with a dense protective webbing

Defoliation occurs in heavy infestations with a litter of yellow-spotted green leaves under trees

Olives

The olive is a species of small tree in the family Oleaceae, much of Africa, the Mediterranean Basin from Portugal to Israel, the Arabian Peninsula, and southern Asia as far east as China, plus the Canary Islands, Mauritius and Réunion. The species is cultivated in many places and considered naturalized in France, Corsica, Crimea, Egypt, Iran, Iraq, Syria, Java, Norfolk Island, California and Bermuda.

Its fruit, also called the olive, is of major agricultural importance in the Mediterranean region as the source of olive oil. The tree and its fruit give its name to the plant family, which also includes species such as lilacs, jasmine, *Forsythia* and the true ash trees (*Fraxinus*).

Description

Figure: *19th century illustration*

The olive tree, *Olea europaea*, is an evergreen tree or shrub native to the Mediterranean, Asia and Africa. It is short and squat, and rarely exceeds 8–15 m (26–49 ft) in height. However, the *Pisciottana*, a unique variety comprising 40,000 trees found only in the area around Pisciotta in the Campania region of southern Italy often exceeds this, with correspondingly large trunk diameters. The silvery green leaves are oblong, measuring 4–10 cm (1.6–3.9 in) long and 1–3 cm (0.39–1.18 in) wide. The trunk is typically gnarled and twisted.

The small white, feathery flowers, with ten-cleft calyx and corolla, two stamens and bifid stigma, are borne generally on the previous year's wood, in racemes springing from the axils of the leaves.

The fruit is a small drupe 1–2.5 cm (0.39–0.98 in) long, thinner-fleshed and smaller in wild plants than in orchard cultivars. Olives are harvested in the green to purple stage. Canned black olives may contain chemicals (usually ferrous sulfate) that artificially turn them black. *Olea europaea* contains a seed commonly referred to in American English as a pit or a rock, and in British English as a stone.

Cultivars

There are hundreds of cultivars of the olive tree (*Olea europaea*). An olive's cultivar has a significant impact on its colour, size, shape, and growth characteristics, as well as the qualities of olive oil. Olive cultivars may be used primarily for oil, eating, or both. Olives for consumption are generally referred to as "table olives".

Since many olive cultivars are self-sterile or nearly so, they are generally planted in pairs with a single primary cultivar and a secondary cultivar selected for its ability to fertilize the primary one. In recent times, efforts have been directed at producing hybrid cultivars with qualities such as resistance to disease, quick growth and larger or more consistent crops.

Uses

The olive tree, *Olea europaea*, has been cultivated for olive oil, fine wood, olive leaf, and the olive fruit. 90% of all harvested olives are turned in to oil, while about 10% are used as table olives.

Table Olives

Table olives are classified by the IOC into 3 groups according to the degree of ripeness achieved before harvesting:

1. Green Olives. Picked when they have obtained full size, but before the ripening cycle has begun. Usually shades of green to yellow.

2. Semi-ripe or Turning Colour Olives. Picked at the beginning of the ripening cycle, when the colour has begun to change from green to multi-colour shades of red to brown. Only the skin is coloured as the flesh of the fruit lacks pigmentation at this stage, unlike that of ripe olives.
3. Black Olives or Ripe Olives. Picked at full maturity when fully ripe. Found in assorted shades of purple to brown to black.

Cultivation

Figure: *An example of black, ripe olives*

The earliest evidence for the domestication of olives comes from the Chalcolithic Period archaeological site of Teleilat Ghassul in what is today modern Jordan. Farmers in ancient times believed that olive trees would not grow well if planted more than a certain distance from the sea; Theophrastus gives 300 stadia (55.6 km or 34.5 mi) as the limit. Modern experience does not always confirm this, and, though showing a preference for the coast, they have long been grown further inland in some areas with suitable climates, particularly in the southwestern Mediterranean (Iberia, northwest Africa) where winters are mild.

Figure: *Olives at a market in Toulon, France*

Growth and Propagation

Olive trees, *Olea europaea*, show a marked preference for calcareous soils, flourishing best on limestone slopes and crags, and coastal climate conditions. They grow in any light soil, even on clay if well drained, but in rich soils they are predisposed to disease and produce poorer oil than in poorer soil. (This was noted by Pliny the Elder.) Olives like hot weather and sunny positions without any shade while temperatures below "10 °C (14 °F) may injure even a mature tree. They tolerate drought well, thanks to their sturdy and extensive root system. Olive trees can live for several centuries, and can remain productive for as long if they are pruned correctly and regularly.

In situations where extreme cold has damaged or killed the olive tree the rootstock can survive and produce new shoots which in turn become new trees. In this way olive trees can regenerate themselves. In Tuscany in 1985 a very severe frost destroyed many productive, and aged, olive trees and ruined many farmers' livelihoods. However new shoots appeared in the spring and, once the dead wood was removed, became the basis for new fruit-producing trees. In this way an olive tree can 'live' for centuries or even millennia.

Olives grow very slowly, and over many years the trunk can attain a considerable diameter. A. P. de Candolle recorded one exceeding 10 m (33 ft) in girth. The trees rarely exceed 15 m (49 ft) in height, and are generally confined to much more limited dimensions by frequent pruning. The yellow or light greenish-brown wood is often finely veined with a darker tint; being very hard and close-grained, it is valued by woodworkers. There are only a handful of olive varieties that can be used to cross-pollinate. Pendolino olive trees are partially self-fertile, but pollenizers are needed for a large fruit crop. Other compatible olive tree pollenizers include Leccino and Maurino. Pendolino olive trees are used extensively as pollenizers in large olive tree groves.

Olives are propagated by various methods. The preferred ways are cuttings and layers; the tree roots easily in favourable soil and throws up suckers from the stump when cut down. However, yields from trees grown from suckers or seeds are poor; they must be budded or grafted onto other specimens to do well (Lewington and Parker, 114). Branches of various thickness cut into lengths of about 1 m (3.3 ft) planted deeply in manured ground soon vegetate. Shorter pieces are sometimes laid horizontally in shallow trenches and, when covered with a few centimetres of soil, rapidly throw up sucker-like shoots. In Greece, grafting the cultivated tree on the wild tree is a common practice. In Italy, embryonic buds, which form small swellings on the stems, are carefully excised and planted under the soil surface, where they soon form a vigorous shoot.

The olive is also sometimes grown from seed. To facilitate germination, the oily pericarp is first softened by slight rotting, or soaked in hot water or in an alkaline solution.

Where the olive is carefully cultivated, as in Languedoc and Provence, the trees are regularly pruned. The pruning preserves the flower-bearing shoots of the preceding year, while keeping the tree low enough to allow the easy gathering of the fruit. The spaces between the trees are regularly fertilized. The crop from old trees is sometimes enormous, but they seldom bear well two years in succession, and in many cases a large harvest occurs every sixth or seventh season.

Old Olive Trees

The olive tree, *Olea europaea*, is very hardy: drought-, disease- and fire-resistant, it can live to a great age. Its root system is robust and capable of regenerating the tree even if the above-ground structure is destroyed. The older the olive tree, the broader and more gnarled the trunk becomes. Many olive trees in the groves around the Mediterranean are said to be hundreds of years old, while an age of 2,000 years is claimed for a number of individual trees; in some cases, this has been scientifically verified.

Pliny the Elder told about a sacred Greek olive tree that was 1,600 years old. An olive tree in west Athens, named “Plato’s Olive Tree”, was said to be a remnant of the grove within which Plato’s Academy was situated, which would make it approximately 2,400 years old. The tree comprised a cavernous trunk from which a few branches were still sprouting in 1975, when a traffic accident caused a bus to fall on and uproot it. Since then, the trunk has been preserved and displayed in the nearby Agricultural University of Athens. A supposedly older tree, the “Peisistratos Tree”, is located by the banks of the Cephisus River, in the municipality of Agioi Anargyroi, and is said to be a remnant of an olive grove that was planted by Athenian tyrant Peisistratos in the 6th century BC. Numerous ancient olive trees also exist near Pelion in Greece. The age of an olive tree in Crete, the Finix Olive is claimed to be over 2,000 years old; this estimate is based on archaeological evidence around the tree.

An olive tree in Algarve, Portugal, is 2000 years old, according to radiocarbon dating.

An olive tree in Bar, Montenegro, is claimed to be over 2,000 years old.

An olive tree on the island of Brijuni (Brioni), Istria in Croatia, has been calculated to be about 1,600 years old. It still gives fruit (about 30 kg or 66 lb per year), which is made into top quality olive oil.

The town of Bshaale, Lebanon claims to have the oldest olive trees in the world (4000 BC for the oldest), but no scientific study supports these claims. Other trees in the towns of Amioun appear to be at least 1,500 years old.

There are dozens of ancient olive trees throughout Israel and Palestine whose age has earlier been estimated to be 1,600–2,000 years old; however, these estimates could not be supported by current scientific practices. Ancient trees include two giant olive trees in Arraba and five trees in Deir Hanna, both in the Galilee region, which have been determined to be over 3,000 years old, although there is no available data to support the credibility of the study that produced these age estimates and as such the 3000 years age estimate can not be considered valid. All seven trees continue to produce olives. Several trees in the Garden of Gethsemane (from the Hebrew words "gat shemanim" or olive press) in Jerusalem are claimed to date back to the purported time of Jesus.

Some Italian olive trees are believed to date back to Roman times, although identifying progenitor trees in ancient sources is difficult. A tree located in *Santu Baltolu di Carana* (municipality of Luras) in Sardinia, Italy, named with respect as the *Ozzastru* by the inhabitants of the region, is claimed to be 3,000 to 4,000 years old according to different studies. There are several other trees of about 1,000 years old within the same garden. The 15th-century trees of *Olivo della Linza* located in Alliste province of Lecce in Puglia were noted by Bishop Ludovico de Pennis during his pastoral visit to the Roman Catholic Diocese of Nardò-Gallipoli in 1452.

Pests, Diseases, and Weather

There are various pathologies that can affect olives. The most serious pest is the olive fruit fly (*Bactrocera oleae*) which lays its eggs in the olive most commonly just before it becomes ripe in the autumn. The region surrounding the puncture rots, becomes brown and takes a bitter taste making the olive unfit for eating or for oil. For controlling the pest the practice has been to spray with insecticides (organophosphates, e.g. dimethoate). Classic organic methods have now been applied such as trapping, applying the bacterium Bacillus thuringiensis and spraying with kaolin. Such methods are obligatory for organic olives.

A fungus, *Cycloconium oleaginum*, can infect the trees for several successive seasons, causing great damage to plantations. A species of bacterium, *Pseudomonas savastanoi* pv. *oleae*, induces tumour growth in the shoots. Certain lepidopterous caterpillars feed on the leaves and flowers.

A pest which spreads through olive trees is the black scale bug, a small black scale insect that resembles a small black spot. They attach themselves firmly to olive trees and reduce the quality of the fruit; their main predators are wasps. The curculio beetle eats the edges of leaves, leaving sawtooth damage.

Rabbits eat the bark of olive trees and can do considerable damage, especially to young trees. If the bark is removed around the entire circumference of a tree it is likely to die. At the northern edge of their cultivation zone, for instance in Southern France and north-central Italy, olive trees suffer occasionally from frost. Gales and long-continued rains during the gathering season also cause damage.

As an Invasive Species

Since its first domestication, *Olea europaea* has been spreading back to the wild from planted groves. Its original wild populations in southern Europe have been largely swamped by feral plants.

In somc other parts of the world where it has been introduced, most notably South Australia, the olive has become a major woody weed that displaces native vegetation. In South Australia, its seeds are spread by the introduced red fox and by many bird species, including the European starling and the native emu, into woodlands, where they germinate and eventually form a dense canopy that prevents regeneration of native trees. As the climate of South Australia is very dry and bushfire prone, the oil rich feral olive tree substantially increases the fire hazard of native sclerophyll woodlands.

Production

Olives are one of the most extensively cultivated fruit crops in the world. In 2011 there were about 9.6 million hectares planted with olive trees, which is more than twice the amount of land devoted to apples, bananas or mangoes. Only coconut trees and oil palms command more space. Cultivation area tripled from 2,600,000 to 7,950,000 hectares (6,400,000 to 19,600,000 acres) between 1960 and 1998 and reached a 10 million ha peak in 2008. The ten largest producing countries, according to the Food and Agriculture Organisation, are all located in the Mediterranean region and produce 95% of the world's olives.

Garlic

Allium sativum, *commonly known as garlic, is a species in the onion genus,* Allium.

Its close relatives include the onion, shallot, leek, chive, and rakkyo. With a history of human use of over 7,000 years, garlic is native to central Asia, and has long been a staple in the Mediterranean region, as well as a frequent seasoning in Asia, Africa, and Europe. It was known to Ancient Egyptians, and has been used for both culinary and medicinal purposes.

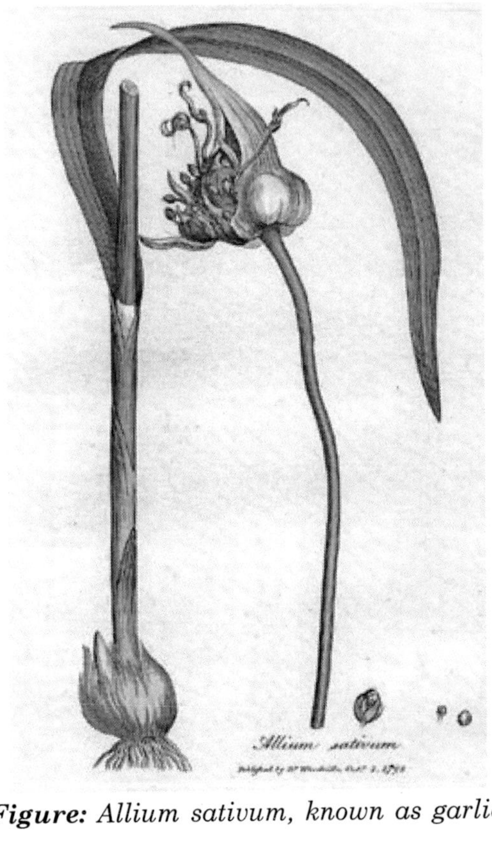

Figure: *Allium sativum, known as garlic,*

Description

Allium sativum *is a bulbous plant. It grows up to 1.2 m (4 ft) in height. Its hardiness is USDA Zone 8. It produces hermaphrodite flowers. Pollination occurs by bees and other insects.*

Origin and Major Types

According to Zohary and Hopf, "A difficulty in the identification of its wild progenitor is the sterility of the cultivars", though it is thought to be descended from the species Allium longicuspis, *which grows wild in central and southwestern Asia.* Allium sativum *grows in the wild in areas where it has become naturalized. The "wild garlic", "crow garlic", and "field garlic" of Britain are members of the species* Allium ursinum, Allium vineale, *and* Allium oleraceum, *respectively. In North America,* Allium vineale *(known as "wild garlic" or "crow garlic") and* Allium canadense, *known as "meadow garlic" or "wild garlic" and "wild onion", are common weeds in fields. One of the best-known "garlics", the so-called elephant garlic, is actually a wild leek (*Allium ampeloprasum*), and not a true garlic. Single clove garlic (also called pearl or solo garlic) originated in the Yunnan province of China.*

European Garlic

Figure: *Flower head*

There are a number of garlics with Protected Geographical Status in Europe; these include:

- Aglio Rosso di Nubia (Red Garlic of Nubia) from Nubia-Paceco, Provincia di Trapani, Sicily, Italy
- Aglio Bianco Polesano from Veneto, Italy (PDO)
- Aglio di Voghiera from Ferrara, Emilia-Romagna, Italy (PDO)
- Ail blanc de Lomagne from Lomagne in the Gascony area of France (PGI)
- Ail de la Drôme from Drôme in France (PGI)
- Ail rose de Lautrec a rose/pink garlic from Lautrec in France (PGI)

- Ajo Morado de las Pedroñeras a rose/pink garlic from Las Pedroñeras in Spain (PGI)

Subspecies and Varieties

There are two subspecies of A. sativum, *ten major groups of varieties, and hundreds of varieties or cultivars.*

- *A. sativum* var. *ophioscorodon* (Link) Döll, called Ophioscorodon, or hard necked garlic, includes porcelain garlics, rocambole garlic, and purple stripe garlics. It is sometimes considered to be a separate species, *Allium ophioscorodon* G.Don.
- *A. sativum* var. *sativum*, or soft-necked garlic, includes artichoke garlic, silverskin garlic, and creole garlic.

Cultivation

Garlic is easy to grow and can be grown year-round in mild climates. While sexual propagation of garlic is indeed possible, nearly all of the garlic in cultivation is propagated asexually, by planting individual cloves in the ground. In cold climates, cloves are planted in the autumn, about six weeks before the soil freezes, and harvested in late spring. The cloves must be planted at sufficient depth to prevent freeze/thaw which causes mold or white rot Garlic plants are usually very hardy, and are not attacked by many pests or diseases. Garlic plants are said to repel rabbits and moles. Two of the major pathogens that attack garlic are nematodes and white rot disease, which remain in the soil indefinitely after the ground has become infected. Garlic also can suffer from pink root, a typically nonfatal disease that stunts the roots and turns them pink or red.

Garlic plants can be grown closely together, leaving enough space for the bulbs to mature, and are easily grown in containers of sufficient depth. Garlic does well in loose, dry, well drained soils in sunny locations, and is hardy throughout USDA climate zones 4–9. When selecting garlic for planting, it is important to pick large bulbs from which to separate cloves. Large cloves, along with proper spacing in the planting bed, will also improve bulb size. Garlic plants prefer to grow in a soil with a high organic material content, but are capable of growing in a wide range of soil conditions and pH levels.

There are different types or subspecies of garlic, most notably hardneck garlic and softneck garlic. The latitude where the garlic is grown affects the choice of type as garlic can be day-length sensitive. Hardneck garlic is generally grown in cooler climates; softneck garlic is generally grown closer to the equator.

Garlic scapes are removed to focus all the garlic's energy into bulb growth. The scapes can be eaten raw or cooked.

Production Trends

Garlic is grown globally, but China is by far the largest producer of garlic, with approximately 10.5 million tonnes (23 billion pounds) grown annually, accounting for over 77% of world output. India (4.1%) and South Korea (2%) follow, with Egypt and Russia (1.6%) tied in fourth place and the United States (where garlic is grown in every state except for Alaska) in sixth place (1.4%). This leaves 16% of global garlic production in countries that each produce less than 2% of global output. Much of the garlic production in the United States is centred in Gilroy, California, which calls itself the "garlic capital of the world".

Uses

Culinary Uses: Garlic is widely used around the world for its pungent flavour as a seasoning or condiment.

Figure: *Garlic being crushed using a garlic press*

Figure: *String of garlic*

The garlic plant's bulb is the most commonly used part of the plant. With the exception of the single clove types, garlic bulbs are normally divided into numerous fleshy sections called cloves. Garlic cloves are used for consumption (raw or cooked) or for medicinal purposes. They have a characteristic pungent, spicy flavour that mellows and sweetens considerably with cooking.

Other parts of the garlic plant are also edible. The leaves and flowers (bulbils) on the head (spathe) are sometimes eaten. They are milder in flavour than the bulbs, and are most often consumed while immature and still tender. Immature garlic is sometimes pulled, rather like a scallion, and sold as "green garlic". When green garlic is allowed to grow past the "scallion" stage, but not permitted to fully mature, it may produce a garlic "round", a bulb like a boiling onion, but not separated into cloves like a mature bulb. Additionally, the immature flower stalks (scapes) of the hardneck and elephant types are sometimes marketed for uses similar to asparagus in stir-fries.

Inedible or rarely eaten parts of the garlic plant include the "skin" and root cluster. The papery, protective layers of "skin" over various parts of the plant are generally discarded during preparation for most culinary uses, though in Korea immature whole heads are sometimes prepared with the tender skins intact. The root cluster attached to the basal plate of the bulb is the only part not typically considered palatable in any form.

Garlic is a fundamental component in many or most dishes of various regions, including eastern Asia, South Asia, Southeast Asia, the Middle East, northern Africa, southern Europe, and parts of South and Central America. The flavour varies in intensity and aroma with the different cooking methods. It is often paired with onion, tomato, or ginger. The parchment-like skin is much like the skin of an onion, and is typically removed before using in raw or cooked form. An alternative is to cut the top off the bulb, coat the cloves by dribbling olive oil (or other oil-based seasoning) over them, and roast them in an oven. Garlic softens and can be extracted from the cloves by squeezing the (root) end of the bulb, or individually by squeezing one end of the clove. In Korea, heads of garlic are fermented at high temperature; the resulting product, called black garlic, is sweet and syrupy, and is now being sold in the United States, United Kingdom and Australia.

Garlic may be applied to different kinds of bread to create a variety of classic dishes, such as garlic bread, garlic toast, bruschetta, crostini and canapé.

Oils can be flavoured with garlic cloves. These infused oils are used to season all categories of vegetables, meats, breads and pasta.

In some cuisines, the young bulbs are pickled for three to six weeks in a mixture of sugar, salt, and spices. In eastern Europe, the shoots are pickled and eaten as an appetizer. Laba garlic, prepared by soaking garlic in vinegar, is a type of pickled garlic served with dumplings in northern China to celebrate the Chinese New Year.

Lightly smoked garlic is becoming increasingly popular in British and European cuisine. It is particularly prized for stuffing poultry and game, and in soups and stews. In both these cases it is important to utilise the undiscarded skin, as much of the smoke flavour is situated there, rather than in the cloves themselves.

Immature scapes are tender and edible. They are also known as "garlic spears", "stems", or "tops". Scapes generally have a milder taste than the cloves. They are often used in stir frying or braised like asparagus. Garlic leaves are a popular vegetable in many parts of Asia. The leaves are cut, cleaned, and then stir-fried with eggs, meat, or vegetables.

Mixing garlic with egg yolks and olive oil produces *aioli*. Garlic, oil, and a chunky base produce *skordalia*. Blending garlic, almond, oil, and soaked bread produces *ajoblanco*. Yogurt mixed with garlic and salt is a common sauce in Eastern Mediterranean cuisines.

Garlic powder has a different taste from fresh garlic. If used as a substitute for fresh garlic, 1/8 teaspoon of garlic powder is equivalent to one clove of garlic.

Storage

Figure: *A basket of garlic bulbs*

Figure: *Ready peeled garlic cloves sold in a plastic container*

Domestically, garlic is stored warm [above 18 °C (64 °F)] and dry to keep it dormant (lest it sprout). It is traditionally hung; softneck varieties are often braided in strands called plaits or *grappes*. Peeled cloves may be stored in wine or vinegar in the refrigerator. Commercially, garlic is stored at 0 °C (32 °F), in a dry, low-humidity environment. Garlic will keep longer if the tops remain attached.

Garlic is often kept in oil to produce flavoured oil; however, the practice requires measures to be taken to prevent the garlic from spoiling. Untreated garlic kept in oil can support the growth of *Clostridium botulinum* which causes the deadly botulism illness; refrigeration will not assure the safety of garlic kept in oil. To reduce this risk, the oil should be refrigerated and used within one week. According to wikihow, the garlic immersed in oil should be stored in the freezer and not the fridge. Commercially prepared oils are widely available. Manufacturers add acids and/or other chemicals to eliminate the risk of botulism in their products. Two outbreaks of botulism related to garlic stored in oil have been reported.

Garlic bulbs should be clean and white with a dried neck and outer skin and quite firm under pressure. They should be discarded if they are soft or spongy or show signs of mould.

In 1961, Chester Lilley from Kent in England was the first person to transform garlic into a pill form for storage. Although not widely accepted at the time for culinary uses, a capsulate solution for both the storage and simple dosing of garlic has become commonplace.

Historical Use

Garlic has been used as both food and medicine in many cultures for thousands of years, dating at least as far back as when the Giza pyramids were built. Garlic is still grown in Egypt, but the Syrian variety is the kind most esteemed now.

Hippocrates, Galen, Pliny the Elder, and Dioscorides all mention the use of garlic for many conditions, including parasites, respiratory problems, poor digestion, and low energy. Its use in China dates back to 2000 BCE.

It was consumed by ancient Greek and Roman soldiers, sailors, and rural classes (Virgil, *Ecologues* ii. 11), and, according to Pliny the Elder (*Natural History* xix. 32), by the African peasantry. Galen eulogizes it as the "rustic's theriac" (cure-all), and Alexander Neckam, a writer of the 12th century, recommends it as a palliative for the heat of the sun in field labour.

In the account of Korea's establishment as a nation, a tiger and a bear prayed to Hwanung that they may become human. Upon hearing their prayers, Hwanung gave them 20 cloves of garlic and a bundle of mugwort, ordering them to eat only this sacred food and remain out of the sunlight for 100 days. The tiger gave up after about twenty days and left the cave. However, the bear remained and was transformed into a woman.

In his *Natural History*, Pliny gives an exceedingly long list of scenarios in which it was considered beneficial (*N.H.* xx. 23). Dr. T. Sydenham valued it as an application in confluent smallpox, and, says Cullen (*Mat. Med.* ii. p. 174, 1789), found some dropsies cured by it alone. Early in the 20th century, it was sometimes used in the treatment of pulmonary tuberculosis or phthisis.

Pests and Plant Protection

Garlic is attacked by many diseases and insects pests. Therefore garlic growers need to know the details of symptoms of diseases and insect pests attacking garlic and their control measures for getting healthy crop. Diseases and insects pests, and their control measures attacking garlic crop are described as under :

Diseases

Purple Blotch: Purple blotch (*Alternaria porri*) appears on leaves as small whitish sunken lesions with purple centres that rapidly enlarge.

The leaves fall over gradually. Spraying of Mancozeb @ 2.5 g/litre of water at 15 days intervals gives good control.

Stemphylium Blight

Stemphylium blight (*Stemphylium vesicarium*) appears as small, yellow to orange flecks or streaks on leaf. These soon develop into elongated, spindle-shaped to ovate elongate, diffused spots, often reaching the leaf tips. They usually turn gray at the centre, brown to dark olive brown with the development of conidiophores and conidia of the pathogen. The spots frequently coalesce into extended patches blightening the leaves and gradually the entire foliage.

Cercospora Leaf Blight

Caused by *Cercospora duddiae*,this disease appears on leaves as small, ash-coloured and irregular shaped spots scattered on leaf lamina. The spots coalesce gradually and results in blightening of foliage. Spraying of ziram or captan @ 2.0 g/litre of water or copper oxychloride @ 3.0 g /litre of water at fortnightly intervals gives good control.

Powdery Mildew

Powdery mildew is caused by *Leveillula taurica*. Distinct pale-yellow patches of variable size on abaxial surface of leaf associated with powdery mass are its main characteristics. Sulphur fungicides @ 2.0 g/ litre of water if sprayed at regular intervals of 15 days after disease appearance gives good control.

Mosaic Disease

Garlic plants infected with mosaic virus show typical symptoms of chlorotic mottling and strips on first emerging leaf followed by pale-yellow broken stripes, resulting in typical mosaic pattern on matured leaves. Yellowish dots on leaves, whitish leaf margin or twisting of leaves are also recorded on a few cultivars. Generally symptoms are mild on younger leaves than on matured leaves. Bulbs harvested from mosaic affected plants after maturity remain smaller in size and cloves are fewer in number. Since virus is transmitted through aphids, spraying of monocrotophos @ 0.5 ml/litre of water or methyl demeton @ 0.75 ml/ litre of water is useful.

Insect Pests

Thrips: Thrips (*Thrips tabaci*) are perhaps the most important insect pests attacking garlic. They are most common during warm weather. They feed on leaf surfaces, causing them to whiten or silver. They are slender about 1/25" long usually hide in angles of leaves.

Spraying of Malathion @ 1 ml/ litre or methyldemeton @ 0.75 ml/ litre of water gives good control if sprayed at frequent intervals. Four sprays of fluvalinate (0.04%) at fortnightly interval at Karnal and malathion (0.1%) at Nasik gave best performance in controlling thrips. Phorate @ 1 kg ai/ha at planting and 30 days after planting controls thrips effectively.

Stem and Bulb Nematode

Caused by *Ditylenchus dipsaci*, it is a tiny worm almost invisible to the naked eye which invades the tissues of garlic, and can seriously reduce the yield. The infestation causes the basal portion of mature plants to swell and become spongy and frequently broken by longitudinal splits. The tissue rots, the leaves and stems become rotten and twisted and the plant is stunted. In severely infested plants, basal plants may be so rotted that it pulls away when the plant is lifted, leaving the roots in the ground. The nematodes are, however, not common in garlic in India.

Pre-planting soil fumigation with dichloro-propene-dichloropropane mixture @ 500 kg/ha is recommended. Ethylene dibromide and dibromochloropropane (Nemagon) should not be used as these leave residues in soil.

Mites

The eriophyid mite, *Aceria tulipae*, a common pest of garlic is so small that its infestations are frequently over-looked. Young garlic plants injured by mites usually have destroyed and twisted leaves with conspicuous yellow or ligat green streaking. The leaves may not emerge readily from the cloves and leaf blades separate poorly after emergence. Bulbs stored for long period may be severely attacked by mites. The cloves wither and bulbs lose their firmness.

Rhizoglyphus Callae and Glycohaqus Domesticus

These may attack on bulbs in storage. Fumigation of dry bulbs with methyl bromide after harvesting and before storage control the mites. Application of methyl bromide @ 1kg/1000 cubic feet for 2 hours at 26°C is recommended.

Physiological Disorders

Sprouting of bulbs in the field is noticed sometimes towards the start of maturity stage of bulbs particularly when there are winter rains or excessive soil moisture and nitrogen supply. This disorder is, however, not of permanent nature and varies from variety-to-variety. Early-planting also causes sprouting. Splitting is also noticed sometimes in some varieties, which is due to delayed harvesting or irrigation after long spell of drought. For past few years rubberification problem is noticed in Rajkot area and also Nilgiri hills of Tamil Nadu. It is increasing day-by-day.

The actual cause for the disorder was since not identified, a trial was planned in collaboration with TNAU at HRS, Ooty, during 1995-96. The effects of insecticides, fungicides, micronutrients and growth regulator were studied. It was observed that rubberification was totally controlled by application of micronutrients i.e. zinc sulphate and ammonium molybdate. It was also controlled by neem cake insecticides and growth regulator like GA.

Further studies are required to eliminate this problem in different areas as the incidence is noticed variably in different areas.

The aerial bulbil formation is also quite common in garlic where lower temperature prevails for more period or there are more temperature variation. It, however, does not affect much on yield or quality of bulbs and varietal variation is noticed where G-282 is seen more prone for this type of effect.

The rubberification and premature sprouting of bulbs are main physiological disorders in garlic. The reasons for these disorders are summarized as under for hill grown garlic in Tamil Nadu.

- Rubberification and premature sprouting of bulbs are noticed mostly in fields which are located in low-lying areas of watershed where there is heavy deposition of nutrients alongwith silt during heavy rains.
- These problems are severe in garlic fields which are more frequently irrigated than the normal requirements of garlic bulbs.
- With the application of higher levels of nitrogen, there is an increased level of pre-mature sprouting of bulbs which results in splitting and rubberification of bulbs.
- Rubberification incidence also increases when higher level of nitrogen is applied in the form of urea.
- Due to increased levels of nitrogen, in addition to the increased production of rubberized, there is an increased level of thrips incidence in leaves during the later stage of crop. The thrips lacerate leaves and cause severe damage to the crop. Through lacerated wounds by thrips, there is a severe incidence of blast disease, caused by *Boltrytis allii,* when crops is 70-90 days old, which further deteriorates the quality of garlic bulbs.

- Short duration type of garlic (Mettupalayam type) is more susceptible than long duration type (Singapore type).
- Premature sprouting of bulbs is more in the crop (April-August) when there is high moisture level in soil at maturity of crop due to heavy rains.
- Delayed harvest during rainy season has increased premature sprouting and splitting of bulbs.
- Wider spacing of cloves at the time of planting increases uptake of nitrogen and water by the individual plants which increases premature sprouting and rubbenfication of garlic bulbs.

Chillies

The chili (also chile pepper or chilli pepper, from Nahuatl *chîlli* is the fruit of plants from the genus *Capsicum*, members of the nightshade family, Solanaceae. The term in British English and in Australia, New Zealand, India, Malaysia and other Asian countries is just chilli without "pepper".

The substances that give chili peppers their intensity when ingested or applied topically are capsaicin (8-methyl-*N*-vanillyl-6-nonenamide) and several related chemicals, collectively called *capsaicinoids*.

Chili peppers originated in the Americas. After the Columbian Exchange, many cultivars of chili pepper spread across the world, used in both food and medicine. Chilies were brought to Asia by Portuguese navigators during the 16th century.

India is the world's largest producer, consumer and exporter of chili peppers. Guntur in Andhra Pradesh produces 30% of all the chilies produced in India, and the state of Andhra Pradesh as a whole contributes 75% of India's chili exports.

Species and Cultivars

Figure: *Thai pepper, similar in variety to the African birdseye, exhibits considerable strength for its size*

The five domesticated species of chili peppers are as follows:

- *Capsicum annuum*, which includes many common varieties such as bell peppers, wax, cayenne, jalapeños, and the chiltepin
- *Capsicum frutescens*, which includes malagueta, tabasco and Thai peppers, piri piri, and Malawian Kambuzi
- *Capsicum chinense*, which includes the hottest peppers such as the naga, habanero, Datil and Scotch bonnet
- *Capsicum pubescens*, which includes the South American rocoto peppers
- *Capsicum baccatum*, which includes the South American aji peppers

Figure: *Immature chilis in the field*

Though there are only a few commonly used species, there are many cultivars and methods of preparing chili peppers that have different names for culinary use. Green and red bell peppers, for example, are the same cultivar of *C. annuum*, immature peppers being green. In the same species are the jalapeño, the poblano (which when dried is referred to as ancho), New Mexico (which is also known as chile colorado), Anaheim, serrano, and other cultivars.

Peppers are commonly broken down into three groupings: bell peppers, sweet peppers, and hot peppers. Most popular pepper varieties are seen as falling into one of these categories or as a cross between them.

Common Pests and Their Control

Bacterial spot: *Xanthomonas vesicatoria*

Symptoms

Water-soaked lesions that dry out and turn brown forming on the underside of the leave; raised brown cankers on stems; cracked brown lesions on fruit.

Cause: Bacterium

Comments: Disease emergence favoured by warm, wet conditions

Management

Use disease free planting material; remove and destroy all crop debris after harvest, or plow material deeply under soil.

Damping-off *Pythium* spp.

Rhizoctonia solani

Figure: *Shriveled stem of pepper seedling caused by damping-off*

Symptoms: Seeds did not germinate; seedlings collapsing and dying; dark stems which are shriveled near the soil line; water-soaked lesions on the stem; water-soaked discoloured roots.

Cause: Fungi

Comments: Disease can be spread in infected soil and irrigation water; disease emergence favoured by overcrowded plants and excessive nitrogen fertilization.

Management: Avoid planting in poorly draining, cool, wet soil; planting in raised beds will help with soil drainage; plant high quality seed that germinates quickly; treat seeds with fungicide prior to planting to eliminate fungal pathogens.

Powdery mildew *Leveillula taurica*

Symptoms: White, powdery growth on underside of leaves which starts in patches but starts to spread across whole leaves, including the top surface; yellow-brown discolouration on underside of leaves; leaves curling.

Cause: Fungus

Comments: Disease can occur in both humid and dry conditions; disease spreads most rapidly in humid conditions; disease commonly affects older leaves.

Management: Apply appropriate fungicide if disease is severe.

Mosaic Cucumber mosaic virus (CMV)

Potato virus Y (PVY)

Symptoms: Alternating light and green areas or mosaic patterns on the leaves; leaves distorted and curled; slow plant growth.

Cause: Virus

Comments: Transmitted by aphids; weeds can act as a source of inoculum; insecticides are not effective due to the speed with which aphids can transmit the virus.

Management: Remove weeds around crop; use reflective mulches to deter aphids from plants.

Fusarium wilt *Fusarium oxysporum* most likely .

Symptoms: Yellowing of foliage and wilting upper leaves; wilting spread to all parts of plant; leaves remain attached to plant and are dark green in colour; red-brown discolouration of vascular tissue; plant death.

Cause: Fungus

Comments: Disease emergence favours high soil moisture content

Management: Plant pepper in well draining soils and avoid waterlogging.

Phytophthora blight *Phytophthora caspsici* .

Symptoms: Black lesions on stems; wilting plant; circular gray-brown lesions on leaves; dark lesions on fruit which may be covered in white sporangia.

Cause: Oomycete

Comments: Disease highly water dependent and spores can be spread by water splash; disease can move root to root to infect a row of plants.

Management: Plant disease free transplants; manage water carefully as pathogen depends on water for survival; rotate crops; apply appropriate fungicide.

Southern blight *Sclerotium rolfsii*

Symptoms: Sudden wilting of leaves; yellowing foliage; browning stem above and below soil; browning branches; stem may be covered with fan-like mycelial mat.

Cause: Fungus

Comments: Fungus can survive in soil for long periods; disease emergence favoured by high temperatures, high humidity and acidic soil; disease found mainly in tropical and subtropical regions, including the southern United States.

Management: Remove infected plants; avoid overcrowding plants to promote air circulation; rotate crops with less susceptible plants; plow crop debris deep into soil; provide a barrier to infection by wrapping lower stems of plant with aluminum foil covering below ground portion of stem and 2-3 in above soil line.

Beet Armyworm: *Spodoptera exigua*

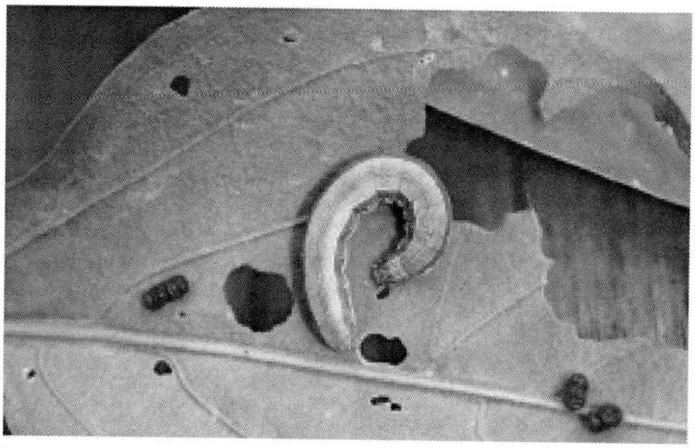

Figure: Late stage beet armyworm larva feeding on pepper leaf

Symptoms: Singular, or closely grouped circular to irregularly shaped holes in foliage; heavy feeding by young larvae leads to skeletonized leaves; shallow, dry wounds on fruit; egg clusters of 50-150 eggs may be present on the leaves; egg clusters are covered in a whitish scale which gives the cluster a cottony or fuzzy appearance; young larvae are pale green to yellow in colour while older larvae are generally darker green with a dark and light line running along the side of their body and a pink or yellow underside.

Cause: Insect

Comments: Insect can go through 3–5 generations a year

Management: Organic methods of controlling armyworms include biological control by natural enemies which parasitize the larvae and the application of *Bacillus thuringiensis*; there are chemicals available for commercial control but many that are available for the home garden do not provide adequate control of the larvae.

Flea Beetles Various

Symptoms: Small holes or pits in leaves that give the foliage a characteristic "shothole" appearance; young plants and seedlings are particularly susceptible; plant growth may be reduced; if damage is severe the plant may be killed; the pest responsible for the damage is a small (1.5–3.0 mm) dark coloured beetle which jumps when disturbed; the beetles are often shiny and metallic in appearance.

Cause: Insects

Comments: Younger plants are more susceptible to flea beetle damage than older ones; older plants can tolerate infestation; flea beetles may overwinter on nearby weed species, in plant debris or in the soil; insects may go through a second or third generation in one year.

Management: In areas where flea beetles are a problem, floating row covers may have to be used prior to the emergence of the beetles to provide a physical barrier to protect young plants; plant seeds early to allow establishment before the beetles become a problem - mature plants are less susceptible to damage; trap crops may provide a measure of control - cruciferous plants are best; application of a thick layer of mulch may help prevent beetles reaching surface; application on diamotecoeus earth or oils such as neem oil are effective control methods for organic growers; application of insecticides containing carbaryl, spinosad, bifenthrin and permethrin can provide adequate control of beetles for up to a week but will need reapplied.

Leafminers: *Lyriomyza* spp.

Symptoms: Thin, white, winding trails on leaves; heavy mining can result in white blotches on leaves and leaves dropping from the plant prematurely; early infestation can cause fruit yield to be reduced; adult leafminer is a small black and yellow fly which lays its eggs in the leaf; larave hatch and feed on leaf interior.

Cause: Insects

Comments: Mature larvae drop from leaves into soil to pupate; entire lifecycle can take as little as 2 weeks in warm weather; insect may go through 7 to 10 generations per year.

Management: Check transplants for signs of leafminer damage prior to planting; remove plants from soil immediately after harvest; only use insecticides when leafminer damage has been identified as unnecessary spraying will also reduce populations of their natural enemies.

Aphids: *Myzus persicae*

Symptoms: Small soft bodied insects on underside of leaves and/or stems of plant; usually green or yellow in colour, but may be pink, brown, red or black depending on species and host plant; if aphid infestation is heavy it may cause leaves to yellow and/or distorted, necrotic spots on leaves and/or stunted shoots; aphids secrete a sticky, sugary substance called honeydew which encourages the growth of sooty mold on the plants.

Cause: Insect

Comments

Distinguishing features include the presence of cornicles (tubular structures) which project backwards from the body of the aphid; will generally not move very quickly when disturbed.

Management

If aphid population is limited to just a few leaves or shoots then the infestation can be pruned out to provide control; check transplants for aphids before planting; use tolerant varieties if available; reflective mulches such as silver coloured plastic can deter aphids from feeding on plants; sturdy plants can be sprayed with a strong jet of water to knock aphids from leaves; insecticides are generally only required to treat aphids if the infestation is very high - plants generally tolerate low and medium level infestation; insecticidal soaps or oils such as neem or canola oil are usually the best method of control; always check the labels of the products for specific usage guidelines prior to use.

Tomato Fruit Worm (Corn Earworm): *Helicoverpa zea*

Symptoms

Larvae burrowing into fruit near stem end and feeding on fruit interior causing decay; fruit turn red prematurely and fall from plant causing reduced yield; young caterpillars are cream-white in colour with a black head and black hairs; older larvae may be yellow-green to almost black in colour with fine white lines along their body and black spots at the base of hairs; eggs are laid singly on both upper and lower leaf surfaces and are initially creamy white but develop a brown-red ring after 24 hours and darken prior to hatching.

Cause: Insect

Comments

Adult insect is a pale green to tan, medium sized moth; can be one of the most damaging pests of tomato; insect overwinters as pupae in the soil; the insect is also a damaging pest of corn and is also referred to as the corn earworm

Management

Early treatment of tomato fruitworms is necessary as once they enter the fruit they are protected from sprays; monitor plants for eggs and young larvae and also natural enemies that could be damaged by chemicals; *Bacillus thuringiensis* or Entrust SC may be applied to control insects on organically grown plants; appropriate chemical treatment may be required for control in commercial plantations

Leafroller (Omnivorous Leafroller): *Platynota stultana*

Symptoms

Leaves of plant rolled and tied together with silk webbing; feeding damage to rolled leaves; defoliation of plant; silk webbing may also be present on fruits and fruits may have substantial scarring from feeding damage; larvae wriggle vigorously when disturbed and may drop from plant on a silken thread

Cause: Insect

Comments

Adult insect is a moth which can fly over several miles to find suitable hosts; alfalfa and sugar beet are good hosts

Management

Monitor plants regularly for signs of infestation; remove weeds from plant bases as they can act as hosts for leafrollers; avoid planting pepper in areas where sugarbeet or alfalfa are grown nearby; *Bacillus thuringiensis* or Entrust SC may be applied to control insects on organically grown plants; apply sprays carefully to ensure that treatment reaches inside rolled leaves

Pepper Weevil: *Anthonomus eugenii*

Symptoms

Feeding damage to foliage, buds and tender young fruit pods; larvae feed within buds and fruit; older fruit misshapen and discoloured; buds and fruit may drop from plant; adult weevil is a small black beetle; larvae are small cream-white grubs with brown heads which can reach 0.6 cm (0.25 in) in length when mature

Cause: Insect

Comments

Weevil will go through multiple generations per year; pepper preferred host but weevil will feed on other plants in the nightshade family also

Management

Remove any nightshade plants growing in close proximity to pepper; immediately remove any dropped fruit from soil surfacc; if pepper weevil becomes problematic, rotate crop the following year; organically acceptable control methods include sprays of pyrethrin; destroy pepper crop residue immediately after harvest

Thrips (Western Flower Thrips, Onion Thrips, etc.): *Frankliniella occidentalis*

Thrips tabaci

Figure: Thrips infestation in a pepper blossom

Symptoms

If population is high leaves and buds may be distorted; leaves appear silvery and are speckled with black feces; insects will feed on and damage flowers; most damage occurs through the transmission of Tomato spotted wilt virus (TSWV); insect is small (1.5 mm) and slender and best viewed using a hand lens; adult thrips are pale yellow to light brown and the nymphs are smaller and lighter in colour

Cause: Insect

Comments

Insect transmits Tomato spotted wilt virus; once acquired, the insect retains the ability to transmit the virus for the remainder of its life.

Management

Avoid planting next to onions, garlic or cereals where very large numbers of thrips can build up; use reflective mulches early in growing season to deter thrips; apply appropriate insecticide if thrips become problematic

Spider Mites (Two Spotted Spider Mite): *Tetranychus urticae*

Symptoms

Leaves stippled with yellow; leaves may appear bronzed; webbing covering leaves; mites may be visible as tiny moving dots on the webs or underside of leaves, best viewed using a hand lens; usually not spotted until there are visible symptoms on the plant; leaves turn yellow and may drop from plant

Cause: Arachnid

Comments: Spider mites thrive in dusty conditions; water-stressed plants are more susceptible to attack

Management

In the home garden, spraying plants with a strong jet of water can help reduce buildup of spider mite populations; if mites become problematic apply insecticidal soap to plants; certain chemical insecticides may actually increase mite populations by killing off natural enemies and promoting mite reproduction.

Organic Strategies for Managing Pests

1. ***Learn to Tolerate Some Damage:*** Most healthy Chile plants can tolerate some damage without suffering serious long-term problems or yield reduction. Munched leaves/ damage pods can easily be removed to maintain the attractive appearance of your plant.
2. ***Introduce the 'Good Guys':*** Aphids feeding in the spring can alarm many Chile growers. Introducing natural predators such as Ladybirds, Parasitic wasps and Lacewings will help clean up most local infestations in a month or so.
3. ***Hand Pick/Hunt Down:*** Hunting down snails and slugs and 'disposing of them' can be a highly satisfying exercise particularly if the little blighters have already struck your prized plants.

Night time 'slug hunts' during wet weather can be particularly productive.

4. ***The Water Hose:*** A strong water hose will temporarily dislodge flies, aphids and other pests from mature plants. However be careful not to saturate or damage your plants and avoid this using method on young seedlings.
5. ***Remove Diseased Plants or Plant Parts:*** Simply removing and disposing of badly damaged plants can help reduce the problem and prevent is spreading to adjacent plants.
6. ***Crop Rotation:*** This is particularly important strategy for tackling soil borne pathogens such as Verticillium Wilt and root rot.
7. ***Grow Pest Resistant & Pest Tolerant Plants:*** Many hybridised varieties, particularly some of the newer sweet pepper varieties have been developed to give specific resistance to diseases such as Tobacco Mosaic Virus (TMV) and Bacterial Leaf Spot.
8. ***Innoculate:*** When growing in pots it is likely that sterilized soil has been used. Sterile soil is ripe for colinisation by many forms of bacteria, fungi, & insects. It is quite likely that the first colinization will not be beneficial. However, just as you can buy yogurts containing beneficial bacteria from the supermarket, you can also buy beneficial bacteria for your soil (though it is a little bit more difficult to get hold of). Beneficial Mycorhizzal fungi is also available, and is starting to become popular in many on-line shops. It may also be useful, depending upon the scope of your growing conditions, to introduce beneficial soil dwelling predatory insects. Introducing your own symbiotic bacteria, fungi, and insects limits the likelihood of colonisation by parasitic forms. In addition to aiding growth of the chile plant, & providing tolerance to environmental stresses, many forms of bacteria and Mycorhizzal fungi are also thought to innoculate the chile plant from diseases and viruses. In addition, they are helpful at reducing the conditions that make these diseases and viruses possible.

7

Taro (Arum), Yams and Drumsticks

Araceae is a family of monocotyledonous flowering plants in which flowers are borne on a type of inflorescence called a spadix. The spadix is usually accompanied by, and sometimes partially enclosed in, a spathe or leaf-like bract.

Figure: *Flower of* Xanthosoma sagittifolium

Also known as the arum family, members are often colloquially known as aroids. This family of 107 genera and over 3700 species is most diverse in the New World tropics, although also distributed in the Old World tropics and north temperate regions. A comprehensive genomic study of *Spirodela polyrhiza* was published in February 2014.

The largest collection of living Araceae is maintained at Missouri Botanical Gardens. Another large collection of living Araceae can be found at Munich Botanical Garden, thanks to the efforts of the researcher and aroid authority Josef Bogner.

Description

Species in Araceae are often rhizomatous or tuberous and are often found to contain calcium oxalate crystals or raphides. The leaves can vary considerably from species to species.

The inflorescence is composed of a spadix, which is almost always surrounded by a modified leaf called a spathe. In monoecious aroids (possessing separate male and female flowers, but with both flowers present on one plant), the spadix is usually organised with female flowers towards the bottom and male flowers towards the top. In aroids with perfect flowers, the stigma is no longer receptive when the pollen is released, thus preventing self-fertilization. Some species are dioecious.

Many plants in this family are thermogenic (heat-producing). Their flowers can reach up to 45°C even when the surrounding air temperature is much lower. One reason for this unusually high temperature is to attract insects (usually beetles) to pollinate the plant, rewarding the beetles with heat energy. Another reason is to prevent tissue damage in cold regions.

Some examples of thermogenic Araceae are: Symplocarpus foetidus (eastern skunk cabbage), Amorphophallus titanum (titan arum), Amorphophallus paeoniifolius (elephant foot yam), Helicodiceros muscivorus (dead horse arum lily) and Sauromatum venosum (voodoo lily). Species such as titan arum and the dead horse arum give off a very pungent smell, often resembling a rotten animal, to attract flies to pollinate the plant. The heat produced by the plant helps to convey the scent further.

Taxonomy

Classification

One of the earliest observations of species in Araceae was conducted by Theophrastus in his work *Enquiry into Plants*. It was not until the 16th century that Araceae was recognised as a distinct group of plants. In 1789 Antoine Laurent de Jussieu classified all climbing aroids as

Pothos and all terrestrial aroids as either *Arum* or *Dracontium* in his book *Familles des Plantes*.

The first major system of classification for the family was produced by Heinrich Wilhelm Schott, who published *Genera Aroidearum* in 1858 and *Prodromus Systematis Aroidearum* in 1860. Schott's system was based on floral characteristics, and used a narrow conception of a genus. Adolf Engler produced a classification in 1876, which was steadily refined up to 1920. His system is significantly different from Schott's, being based more on vegetative characters and anatomy.

The two systems were to some extent rivals, with Engler's having more adherents before the advent of molecular phylogenetics brought new approaches.

Modern studies based on gene sequences show the Araceae (including the Lemnoideae, duckweeds) to be monophyletic, and the first diverging group within the Alismatales. The APG III system of 2009 recognises the family, including the genera formerly segregated in the Lemnaceae. The sinking of Lemnaceae into Araceae is not universally accepted. For example, the 2010 *New Flora of the British Isles* uses a paraphyletic Araceae and a separate Lemnaceae.

Genera

Figure: *The cuckoo-pint or lords and ladies (*Arum maculatum*) is a common arum in British woodlands*

Figure: Arisaema triphyllum

Anthurium and *Zantedeschia* are two well-known members of this family, as are *Colocasia esculenta* (taro) and *Xanthosoma roseum* (elephant ear or 'ape). The largest unbranched inflorescence in the world is that of the arum *Amorphophallus titanum* (titan arum). The family includes many ornamental plants: *Dieffenbachia*, *Aglaonema*, *Caladium*, *Nephthytis*, and *Epipremnum*, to name a few. In the genus *Cryptocoryne* are many popular aquarium plants.

Both Colocasia esculenta (Taro) and *Monstera deliciosa* provide food value; the fruit of *M. deliciosa* is called "Mexican breadfruit". *Philodendron* is an important plant in the ecosystems of the rainforests and is often used in home and interior decorating. *Symplocarpus foetidus* (skunk cabbage) is a common North American species. An interesting peculiarity is that this family includes the largest unbranched inflorescence, that of the titan arum, often erroneously called the "largest flower" and the smallest flowering plant and smallest fruit, found in the duckweed, *Wolffia*.

Diseases and Pests

Taro production in the Asia/Pacific region is currently under the stranglehold of one pest (the taro beetle) and one disease (the taro leaf blight), both of which are proving to be extremely menacing to the taro industry. Some countries such as Fiji have only the beetle; others such as Samoa have only the leaf blight; others such Papua New Guinea have both; while yet others such as Tonga have so far escaped either of these two afflictions. In most places where they have occurred, the beetle and the blight have posed serious problems for the taro industry. Their presence has resulted in quarantine isolation for some of the affected countries, and their resultant exclusion from the export taro trade. The most dramatic recent example has been Samoa, where the appearance of the taro leaf blight since 1993 has not only wiped out the lucrative taro export trade, but also seriously destabilised the internal food supply.

A third, but slightly less menacing, affliction of taro is the alomae/ bobone virus disease complex. The characteristics of this and other diseases and pests of taro will be described in this chapter. Their presence and impact on the taro industry in various countries will be taken up in the sections where taro production is discussed for each country.

The Taro Beetle

The taro beetles of economic importance are several species belonging to the genus Papuana *(Coleoptera: Scarabaeidae). These include* Papuana woodlarkiana, Papuana biroi, Papuana huebneri, *and* Papuana trinodosa. *The adult beetle is black, shiny, and 15-20 mm in length. Many species have a horn on the head.*

The adult beetles fly from the breeding sites to the taro field and tunnel into the soil just at the base of the taro corm. They then proceed to feed on the growing corm, leaving large holes that degrade the eventual market quality of the corm. Also the wounds that they create while feeding promote the attack of rot-causing organisms. The feeding activity can cause wilting and even death of the affected plants. After feeding for about two months, the female beetle flies to neighbouring bushes to lay eggs. The eggs are laid 5-15 cm beneath the soil close to a host plant, (Jackson, 1980).

The eggs are cylindrical and brown or white in colour. A wide range of plants have been found to be hosts for taro beetle breeding (Sar et al., 1997). These include Johnson grass (Sorghum verticilliflorum), Elephant grass (Pennisetum purpureum), Kunai (Imperata cylindrica) and pitpit (Phragmites karka). Larvae hatch from the eggs in 11-16 days. The larvae feed on plant roots and dead organic matter at the base of the host plants.

The larva moults about three times in its 3-4 months of life, and then pupates. After about two weeks, the adults develop from the pupa and fly to neighbouring taro plots to cause another cycle of damage. The adult lives for 4-8 months.

Not only does the taro beetle have a wide host range for breeding, but it also has a wide host range of crops that the adult feeds on and disfigures. Crops that are attacked include tannia, sugarcane, banana, sweet potato, yams, etc. This versatility of hosts makes the taro beetle additionally destructive, and its control much more difficult.

The taro beetle is a pest in taro production in a wide sweep of territory from Indonesia through Papua New Guinea, Solomon Islands, Vanuatu to Fiji and New Caledonia; in short, virtually all of Melanesia and beyond. It was first reported in Fiji in 1984.

Numerous efforts have been made to develop effective control measures for the taro beetle. Mulching with polythene, coconut husk or grass has only been partially effective. The earlier recommendation of lindane for taro beetle control in Papua New Guinea has proved to be environmentally unsustainable. Other insecticides have proved not to be effective; nor has the use of physical barriers such as fly wire or shade cloth spread over the soil.

The most recent research efforts are now concentrating on finding an effective biological control. Certain pathogens of the beetle have been identified. These include a fungus (Metarhizium anisopliae), a bacterium (Bacillus popilliae) and the protozoa Vavraia. Much of this research is taking place in Papua New Guinea and Solomon Islands, supported by the Pacific Regional Agricultural Programme (PRAP). Hopefully, a biological control measure for the taro beetle will become available before long.

Taro Leaf Blight

Taro leaf blight is caused by the fungus Phytophthora colocasiae. It was first reported in Java about a century ago, and has since spread to various parts of Asia and the Pacific. The list of countries where it has been reported include Indonesia, Papua New Guinea, Solomon Islands, Hawaii, Samoa, American Samoa, Thailand and the Philippines.

The disease begins as purple-brown water-soaked lesions on the leaf. A clear yellow liquid oozes from the lesions. These lesions then enlarge, join together and eventually destroy the entire lamina in 10-20 days. Free water collecting on older leaves, as well as high temperature and high humidity are conducive to onset and spread of the disease and germination of the spores.

The disease can be spread from plant to plant by wind and splashing rain. Spores survive in planting material for three or more weeks. Thus, infected planting material is one common means of spreading the disease over long distances and from season to season. The disease also attacks Alocasia macrorrhiza, a common aroid crop in the Pacific region, but the symptoms and yield losses are less severe.

The disease can cause yield losses of 30-50%, and results in lowering of the quality of the reduced harvest. Also taro leaves for human consumption are rare in affected areas. Most countries where the disease has been reported are under strict quarantine isolation.

Various approaches have been used to try to control the taro leaf blight. Agronomic methods that have given partial success include careful choice of planting material, planting at high density, intercropping taro with other crops rather than growing it as a sole crop, and crop rotation. Field removal of infected leaves has also been useful, but it is extremely laborious. In Samoa, control has been achieved by an intensive spraying programme with Ridomil or Manzate, and more recently with phosphorous acid (Foschek).

Chemical control is extremely tedious, expensive, and not totally effective. An integrated control approach combining cultural and chemical methods seems to be the best at present. The ultimate solution must lie in the breeding and release of resistant cultivars. The taro breeding programme in Bubia, Papua New Guinea, has already identified several promising lines in this regard.

In some countries/territories, the taro leaf blight is present but causes relatively minor economic damage. This is true of the Philippines, Thailand and Hawaii. In other cases such as Samoa and American Samoa, the disease can be devastating. This situation has led to conjectures about the possibility that various strains of Phytophthora colocasiae may exist, and that in south-east Asia in particular, some of these strains may have evolved along with the taro crop (Lebot, personal communication) and may be less virulent. This factor is in addition to differences in the genetic make-up and genetic diversity of the taro crop in each country.

The Alomae/Bobone Virus Disease Complex

The alomae virus disease is caused by a complex of two or more viruses acting together. The two viruses that are definitely involved are the taro large bacilliform virus (TLBV) which is transmitted by the plant hopper Tarophagus proserpina, and the taro small bacilliform virus (TSBV) which is transmitted by the mealybug Planococus citri (Rodoni 1995). Neither virus is transmissible by mechanical contact, and their host range seems limited to aroids only. The full-blown alomae disease occurs when these two viruses (and possibly others) are present. Presence of only TLBV alone results in bobone, a milder form of the disease.

Alomae first starts as a feathery mosaic on the leaves. Lamina and veins become thick. The young leaves are crinkly and do not unfold normally. The petiole is short and manifests irregular outgrowths on its surface. The entire plant is stunted and ultimately dies. The symptoms of bobone are similar, but the leaves are more stunted and the lamina is curled up and twisted. With bobone, complete death of the entire plant does not usually occur.

Severe cases of alomae can result in total crop loss, while bobone can cause up to 25% yield loss. However, in many instances, only isolated plants in taro fields seem to be affected by either disease, and in the case of bobone infected plants may recover from the symptoms. The alomae/bobone disease complex has been reported in Papua New Guinea and Solomon Islands. The disease is controlled by pulling out diseased plants in the field, and by careful selection to ensure disease-free planting material. Ultimately, control will have to rely on breeding and disseminating resistant cultivars. Some tolerant cultivars bred through recurrent selection, have been released in Solomon Islands since 1992 (Gunua & Kokoa, 1995).

Dasheen Mosaic Virus Disease (DMV)

While the alomae/bobone disease is mainly confined to the Pacific, the dasheen mosaic virus disease occurs world-wide. Most taro-producing countries in the Asia/Pacific region have the disease. DMV is caused by a stylet-borne, flexuous, rod-shaped virus that is spread by aphids. It is characterized by chlorotic and feathery mosaic patterns on the leaf, distortion of leaves, and stunted plant growth. The disease is not lethal, but yield is depressed. Control is through the use of DMV-free planting material, field sanitation, and quarantine measures.

Other Diseases and Pests

Other diseases and pests of taro include:

a) Corm and root rots caused by the fungi *Pythium* spp and *Phytophthora*.

b) Nematodes.

c) The taro planthopper, *Tarophagus proserpina* which not only transmits virus diseases, but can cause wilting and death of the plant in heavy infestations.

d) Aphids.

e) Taro hornworm which defoliates the plant.

f) Armyworms or cluster caterpillars which can also do extensive damage to the leaves.

While these diseases and pests may be considered minor, they can become quite severe in certain locations or at certain times during the cropping season.

Elephant-Foot

Amorphophallus paeoniifolius, the Elephant foot yam or Whitespot giant arum or Stink lily, is a tropical tuber crop that offers excellent scope for adoption in the tropical countries as a cash crop due to its production potential and popularity as a vegetable in various delicious cuisines.

Figure: *Elephant foot yam*

Elephant foot yam is basically a crop of Southeast Asian origin. It grows in wild form in Sri Lanka, the Philippines, Malaysia, Indonesia, and other Southeast Asian countries. In India it is grown mostly in Bihar West Bengal, Kerala, Karnataka, Andhra Pradesh, Maharashtra and Orissa. In India it is popularly known as oal in Hindi *ol* ("2) in Bengali, *suran* or *jimikand* in Hindi, *senai kizhangu* in Tamil, *suvarna gedde* in Kannada, *chena* (G() in Malayalam, *oluo* in Oriya, and *pulla ganda* in Telugu.

However, in Tonga, teve was seen as the most inferior of all yam species, only to be eaten if nothing else was available. In Bihar its used in Oal curry, Oal bharta or chokha, Pickels and chutney also called Barabar chutney as its has mango, ginger and oal in equal quantity hence named barabar (meaning in equal amount).

In West Bengal, these yams are eaten as fried or as yam curry. The plant body of elephant foot yam is also eaten in West Bengal as green vegetable called Bengali: "ol shaak".

As per Indian Medicinal Plants dictionary by C.P. Khare, published by Springer, the Elephant-foot yam has several medicinal benefits and widely used in Indian medicine including Ayuverda, Siddha and Unani. The corm is prescribed in bronchitis, asthma, abdominal pain, emesis, dysentery, enlargement of spleen, piles, elephantiasis, diseases due to vitiated blood, rheumatic swellings.

Along with other therapeutic applications, The Ayurvedic Pharmacopoeia of India indicates the use of corm in prostatic hyperplasia. (The corm is irritant due to the presence of calcium oxalate.

It can be consumed after it is washed well and boiled in tamarind water or butter milk.)

The corm contains an active diastatic enzyme amylase, betulinic acid, tricontane, lupeol, stigmasterol, betasitosterol and its palmitate and glucose, galactose, rhamnose and xylose.

Intake of 3.6 - 7.2 g of konjac (Amorphophallus konjac tuber) mannan for 90 days reduced the dose of insulin or hypoglycaemic drugs (in human clinical study). (Francis Brinker.) Amorphophallus sylvatius Kunth. is equated with the wild var. of Suurana, used especially in the treatment of piles.

Elephant Foot Yam (Amorphophallus Paeoniifolius)

Elephant foot yam requires fairly long growing season and a rainfall of about 150 cm during the crop period. A well-drained soil of medium texture is suited for this crop.

Season

Corm pieces are normally planted during February-March, before the onset of monsoon.

Variety

Sree Padma: The crop matures in 8-9 months. Cooked tubers are free from acridity.

Sree Athira; First genetically improved variety with very good cooking quality.

Seeds and Sowing

Tuber cut-pieces weighing about 1 kg are ideal for planting. Dip the pieces in cowdung slurry and allow to dry under shade before planting. Nematodes associated with amorphophallus can be controlled by seed material treatment with talc based formulation of *Bacillus macerans* @ 3g (10^6cfu/g) per kg of corms.

After planting, cover the pit with dried leaves or other mulching materials. About 12,000 cut pieces weighing about 12 t are required for planting one hectare. Most of the seed material will germinate within one month after planting.

Mealy bugs usually attack the corm in field and store. Avoid planting corms already infested.

Land Preparation

Dig pits of 60 cm x 60 cm x 45 cm size 90 cm apart. Collect the topsoil to a depth of 15-20 cm separately and fill it after the pits are formed. Apply cowdung or compost at 2-2.5 kg/pit and mix with topsoil.

Minisett Technique for Quality Planting Materials

Minisetts weighing 100 g each can be planted directly in nursery beds or in the main field with the central bud portion facing up at a spacing of 60 cm x 45 cm. A total of 37,000 minisetts/ha is required as against 12,345 setts/ha in the traditional method. Multiplication ratio in elephant foot yamcould be enhanced to 1:15 as against the conventional 1:3 by adopting minisett technique.

After Cultivation

Apply full dose of P_2O_5 and half the dose N and K_2O (N:P_2O_5:K_2O @ 50:50:75 kg ha-1) after forty five days of planting along with intercultivation and weeding. Apply second dose of fertilizers (N and K_2O @ 50:75 kg ha-1) one month after the first application along with intercultivation and earthing up.

Harvesting

The crop will be ready for harvest 8-9 months after planting.

Organic Farming Technology for Elephant Foot Yam

Recommendation

- Raising green manure cowpea (seed rate @ 20 kg ha-1) prior to elephant foot yam and incorporation of green matter at 45-60 days.
- Use of organically produced planting materials.

- Treatment of corm pieces of 500-750 g with slurry containing cowdung, neem cake and *Trichoderma harzianum* (5 g/kg seed) and drying under shade before planting.
- Application of *Trichoderma harzianum* incorporated FYM @ 36 t ha-1 (3 kg/pit) in pits at the time of planting (FYM neem

cake mixture (10:1) inoculated with *Trichoderma harzianum* @ 2.5 kg/tonne of FYM neem cake mixture. Trichoderma can be multiplied in FYM alone but it will take 15 days to form sufficient inoculam as against 7-8 days if neem cake is also used along with FYM). This is effective against collar rot caused by *Sclerotium rolfsii.*

- Application of neem cake @ l.0 t ha-1 (80-85 g/pit) in pits at the time of planting.
- Inter-sowing of green manure cowpea (seed rate @ 20 kg ha-1) between elephant foot yam pits and incorporation of green matter in pits a i 45-60 days. The green matter addition from the 2 green manure crops should be 20-25 t ha^{-1}.
- Application of ash @ 3 t ha-1 (250 g/pit) at the time of incorporation of green maure in pits.

Yams

Figure: *Yams at Brixton market*

Yam is the common name for some plant species in the genus *Dioscorea* (family Dioscoreaceae) that form edible tubers. These are perennial herbaceous vines cultivated for the consumption of their starchy tubers in Africa, Asia, Latin America, the Caribbean and Oceania. There are many cultivars of yam.

Although some varieties of sweet potato (*Ipomoea batatas*) are also called *yam* in parts of the United States and Canada, it is not part of the family Dioscoreaceae; rather it is in the morning glory family Convolvulaceae.

The true yam is a versatile vegetable. It can be barbecued, roasted, fried, grilled, boiled, baked, smoked and when grated it is processed into a dessert recipe. Yams are the staple crop of the Igbo people of Nigeria, in their language it is known as *ji*, and they commemorate it by having yam festivals known as *Iri-ji* or *Iwa-Ji* depending on the dialect.

Yams are a primary agricultural and culturally important commodity in West Africa, where over 95 percent of the world's yam crop is harvested. Yams are still important for survival in these regions. Some varieties of these tubers can be stored up to six months without refrigeration, which makes them a valuable resource for the yearly period of food scarcity at the beginning of the wet season. Yam cultivars are also cropped in other humid tropical countries.

Yam tubers can grow up to 1.5 metres (4.9 feet) in length and weigh up to 70 kilograms (154 lb) and 3 to 6 inches high. The vegetable has a rough skin which is difficult to peel, but which softens after heating. The skins vary in colour from dark brown to light pink. The majority of the vegetable is composed of a much softer substance known as the "meat". This substance ranges in colour from white or yellow to purple or pink in mature yams.

Because of their abundance and importance to survival, yams were highly regarded in Jamaican ceremonies and constitute part of many West African ceremonies. Certain species of yams are a competing phytochemical source. Yams are used as food plants by the larvae of some Lepidoptera species including *Palpifer sordida.*

Description

Sweet potatoes and yam, freshly pulled out of ground. Sweet potatoes and yam varieties come in many natural colours.

Yams are monocots, related to lilies and grasses. Native to Africa and Asia, yams vary in size from that of a small potato to over 60 kilograms. There are over 600 varieties of yams and 95 percent of these crops are grown in Africa.

Figure: *Sweet potatoes*

Differences between True Yam and Sweet Potato "Yam"

Compared to sweet potatoes, yams are starchier and drier. Yams are a monocot (a plant having one embryonic seed leaf) and from the Dioscoreaceae family. Sweet Potatoes are a dicot (a plant having two embryonic seed leaves) and are from the Convolvulacea family. The table below lists some important differences between yam and sweet potato.

Factor	***Sweet Potato***	***Yam***
Plant Family	Morning glory	Yam
Chromosomes	2n=90	2n=20
Flower	Monoecious	Dioecious
Origin	Tropical America (Peru, Ecuador)	West Africa, Asia
Edible part	Storage root	Tuber
Appearance	Smooth, with thin skin	Rough, scaly
Shape	Short, blocky, tapered ends	Long, cylindrical, some with "toes"
Mouth feel	Moist	Dry
Taste	Sweet	Starchy
Beta carotene	Usually high	Usually very low
Propagation	Transplants/vine cuttings	Tuber pieces

Major Cultivated Species

There are many cultivars of yam throughout the humid tropics. The most economically important are discussed below.

Dioscorea Rotundata and D. Cayenensis: *Dioscorea rotundata*, the "white yam", and *Dioscorea cayenensis*, the "yellow yam", are native to Africa. They are the most important cultivated yams.

In the past they were considered two separate species but most taxonomists now regard them as the same species. There are over 200 cultivated varieties between them.

White yam's tuber is roughly cylindrical in shape, the skin is smooth and brown and the flesh usually white and firm. Yellow yam is named after its yellow flesh, a colour caused by the presence of carotenoids. It looks similar to the white yam in outer appearance; its tuber skin is usually a bit firmer and less extensively grooved.

The yellow yam has a longer period of vegetation and a shorter dormancy than white yam.

The Kokoro variety is important in making dried yam chips.They are large plants; the vines can be as long as 10 to 12 metres (33 to 39 ft). The tubers most often weigh about 2.5 to 5 kilograms (5.5 to 11.0 lb) each but can weigh as much as 25 kilograms (55 lb).

After 7 to 12 months growth the tubers are harvested. In Africa most are pounded into a paste to make the traditional dish of "pounded yam" (Kay 1987).

Figure: *Yams produced in Brazil, on sale in Finland.*

D. Alata: *Dioscorea alata*, called "water yam", "winged yam" and "purple yam" (not to be confused with the Okinawan purple "yam", which is a sweet potato), was first cultivated in Southeast Asia.

Although not grown in the same quantities as the African yams, it has the largest distribution world-wide of any cultivated yam, being grown in Asia, the Pacific islands, Africa, and the West Indies (Mignouna 2003).

Even in Africa, the popularity of water yam is second only to white yam. The tuber shape is generally cylindrical, but can vary. Tuber flesh is white and watery in texture.

Figure: *Purple yam freshly harvested and sliced for cross-sectional view.*

In the United States, water yam has become an invasive species in somc Southern states.

In the Philippines it is known as "ube" and is used as an ingredient in many sweet desserts. In Indonesia it is known as "ubi". In Vietnam, it is called *khoai má* and is used mainly as an ingredient for soup. In India, it is known as *ratalu* or *violet yam*. In Hawaii it is known as *uhi*. In Japan it is called *daijo* or *beniimo*.

Uhi was brought to Hawaii by the early Polynesian settlers and became a major crop in the 19th century when the tubers were sold to visiting ships as an easily stored food supply for their voyages (White 2003).

D. Opposite: *Dioscorea opposita*, "Chinese yam", is native to China. The Chinese yam plant is somewhat smaller than the African, with the vines about 3 metres (10 feet) long.

Figure: *Segment of a Japanese mountain yam.*

It is tolerant to frost and can be grown in much cooler conditions than other yams. It is now grown in China, Korea, and Japan.

It was introduced to Europe in the 19th century when the potato crop there was falling victim to disease, and is still grown in France for the Asian food market. The tubers are harvested after about 6 months of growth. Some are eaten right after harvesting and some are used as ingredients for other dishes, including noodles, and for traditional medicines (Kay 1987).

D. Bulbifera: *Dioscorea bulbifera*, the "air potato", is found in both Africa and Asia, with slight differences between those found in each place. It is a large vine, 6 metres (20 ft) or more in length. It produces tubers; however the bulbils which grow at the base of its leaves are the more important food product. They are about the size of potatoes (hence the name "air potato"), weighing from 0.5 to 2 kilograms (1.1 to 4.4 lb).

Figure: *Air potato*

Some varieties can be eaten raw while some require soaking or boiling for detoxification before eating. It is not grown much commercially since the flavour of other yams is preferred by most people. However it is popular in home vegetable gardens because it produces a crop after only four months of growth and continues producing for the life of the vine, as long as two years. Also the bulbils are easy to harvest and cook (Kay 1987).

In 1905 the air potato was introduced to Florida and has since become an invasive species in much of the state. Its rapid growth crowds out native vegetation and is very difficult to remove since it can grow back from the tubers, and new vines can grow from the bulbils even after being cut down or burned (Schultz 1993).

D. Esculenta: *Dioscorea esculenta*, the lesser yam, was one of the first yam species cultivated. It is native to Southeast Asia and is the third most commonly cultivated species there, although it is cultivated very little in other parts of the world. Its vines seldom reach more than 3 metres (10 feet) in length and the tubers are fairly small in most varieties.

Figure: *Wild bitter yam vines*

The tubers are eaten baked, boiled, or fried much like potatoes. Because of the small size of the tubers, mechanical cultivation is possible; which, along with its easy preparation and good flavour, could help the lesser yam to become more popular in the future (Kay 1987).

D. Dumetorum: *Dioscorea dumetorum*, the bitter yam, is popular as a vegetable in parts of West Africa; one reason being that their cultivation requires less labour than other yams. The wild forms are very toxic and are sometimes used to poison animals when mixed with bait. It is said that they have also been used for criminal purposes (Kay 1987).

D. Trifida: *Dioscorea trifida*, the cush-cush yam, is native to the Guyana region of South America and is the most important cultivated New World yam. Since they originated in tropical rain forest conditions their growth cycle is less related to seasonal changes than other yams. Because of their relative ease of cultivation and their good flavour they are considered to have a great potential for increased production (Kay 1987).

Production

Yams are farmed on about 5 million hectares in about 47 countries in tropical and subtropical regions of the world.

Yam crop begins when whole seed tubers or tuber portions are planted into mounds or ridges, at the beginning of the rainy season. The crop yield depends on how and where the setts are planted, sizes of mounds, interplant spacing, provision of stakes for the resultant plants, yam species, and tuber sizes desired at harvest. Small-scale farmers in West and Central Africa often intercrop yams with cereals and vegetables. The seed yams are perishable and bulky to transport. Farmers who do not buy new seed yams, usually set aside up to 30 percent of their harvest for planting the next year. Yam crops face pressure from a range of insect pests, fungal and viral diseases, as well as nematodes.

Yam typically grows for six to ten months and is dormant for two to four months, depending on the species. The growth and dormant phases correspond respectively to the wet season and the dry season. For maximum yield the yam requires a humid tropical environment, with an annual rainfall of over 1500 millimetres distributed uniformly throughout the growing season. White, yellow and water yams typically produce a single large tuber per year, generally weighing 5 to 10 kilograms.

Yam production requires high labour. The crop has low yield per hectare compared to crops such as cassava (manioc) or sweet potato. It is not an efficient food staple given the relatively large amount of planting material that is required and its long growing season. Yam's labour requirement exceeds that of other comparable crops. Yam is also difficult to preserve and store over extended periods of time. The cost per 1,000 calories of yam is four times greater than those of other root and tuber crops.

For these reasons and problems of storing harvested yam, the costs of yam production are high and yam crops are slowly losing ground to cassava and other food staples. Despite these high costs and low nutrient density, when compared to other tubers and roots, low technology yam farming produces the highest amount of food calorie and protein annually per hectare per season, on average. Given this nutritional value of yam and its high cultural acceptance in certain parts of Africa, there is an interest in developing knowledge that can improve yam agriculture.

In 2010, the world harvested 48.7 million tonnes of yam, 95 percent of which was produced in Africa. The biggest yam harvest, globally, was in 2008, when the world produced 54 million metric tonnes of yam.

The world average annual yield of yams was 10.2 tonnes per hectare in 2010. The most productive yam farms in the world were in Colombia, where nationwide average annual yield was 28.3 tonnes per hectare. These are average national yields; certain farms report yields significantly above 30 tonnes per hectare, for example for yellow yam; other farms report yields less than 1 tonnes per hectare.

Despite the high labour requirements and production costs, consumer demand for yam is very high in certain subregions of Africa, making yam cultivation quite profitable to certain farmers.

Toxicity

Figure: *Wild yams in Burkina Faso*

Unlike cassava, most varieties of edible, mature, cultivated yam do not contain toxic compounds. However, there are exceptions.

Bitter compounds tend to accumulate in immature tuber tissues of white and yellow yams. These may be polyphenols or tannin-like compounds.

Wild forms of bitter yams do contain some toxins that taste bitter, and hence are referred to as bitter yam. Bitter yams are not normally eaten except at times of desperation in poor countries and in times of local food scarcity. They are usually detoxified by soaking in a vessel of salt water, in cold or hot fresh water or in a stream. The bitter compounds in these yams are water-soluble alkaloids which, on ingestion, produce severe and distressing symptoms. Severe cases of alkaloid intoxication may prove fatal.

Aerial or potato yams too have anti-nutritional factors. In Asia, detoxification methods, involving water extraction, fermentation and roasting of the grated tuber, are used for bitter cultivars of this yam.

The bitter compounds in yams also known locally as *air potato* include diosbulbin and possibly saponins. These substances are toxic, causing paralysis. Extracts are sometimes used in fishing to immobilize the fish and thus facilitate capture. The community of Zulus use this yam as bait for monkeys, while hunters in Malaysia use it to poison wildlife such as tigers. In Indonesia, an extract of *air potato* is used in the preparation of arrow poison.

Nutritional Value

The protein content and quality of roots and tubers is lower than other food staples. Of all roots and tubers, the protein content of yam and potato is the highest, being approximately 2 percent on a fresh weight basis. Yams, with cassava, provide a much greater proportion of the protein intake in Africa, ranging from 6 percent in East and southern Africa to about 16 percent in humid West Africa.

Yam, like other root crops, is not a good source of essential amino acids. It is rich in phenylalanine and threonine but limiting in the sulphur amino-acids, cystine and methionine and in tryptophan. Yam consuming areas of Africa have a high incidence of kwashiorkor, a serious medical condition in children caused by protein deficiency. Experts emphasize the need to supplement a yam-driven diet with more protein-rich foods in order to support active and healthy growth in infants.

Except for potassium, vitamin B_6 and vitamin C, yam is a food with moderate nutrient density. Yam provides around 110 calories per 100 grams. It has good levels of potassium, manganese, thiamin and dietary fibre, while being low in saturated fat and sodium. Yam generally has a lower glycemic index, about 54% of glucose per 150 gram serving, compared to potato products.

The African yam (Dioscorea sp) contains thiocyanate; it was suggested in one 1986 paper that potentially it is protective against sickle cell anemia (SCA). The paper points out that SCA is known to be relatively rarer in Africans than in the African-American population of the United States, and suggests that the higher thiocyanate content of the African diet explains the high incidence of SCA among African-Americans and its rarity in Africans.

Yam is an important dietary element for Nigerian and West African people. It contributes more than 200 calories per person per day for more than 150 million people in West Africa, and is an important source of income. Yam is an attractive crop in poor farms with limited resources. It is rich in starch, and can be prepared in many ways. It is available all year round, unlike other, unreliable, seasonal crops. These characteristics make yam a preferred food and a culturally important food security crop in some sub-Saharan African countries.

Phytochemical Value

The tubers of certain wild yam, a variant of kokoro yam and other species of Dioscorea, such as Dioscorea nipponica, are a source for the extraction of diosgenin, a steroid sapogenin. The extracted diosgenin is used for the commercial synthesis of cortisone, pregnenolone, progesterone, and other steroid products. Such preparations were used in early combined oral contraceptive pills. The unmodified steroid has estrogenic activity.

Comparison to Other Staple Foods

The following table shows the nutrient content of yam and major staple foods in a raw harvested form. Raw forms, however, aren't edible and can not be digested. These must be sprouted, or prepared and cooked for human consumption.

Propagation

Yams are mainly grown in tropical and sub tropical climates and they do not grow well at temperatures below 22°C (71.6°F). They perform best in well-draining fertile soils with a pH between 5.5 and 6.5 in full sun or part shade. Very wet soils should be avoided as this promotes tuber rot. Yams are propagated from small tubers which should be planted to a depth of 15 cm (6 in) allowing at least 30 cm (12 in) between individual plants and 1.5 m (5 ft) between rows. The soil is often mounded around plants or ridged to aid drainage. It is common practice to stake plants with a 2–4 m (6.6–13.2 ft) support to allow them to climb and ensure that all parts of the plant receive adequate sunlight.

Common Pests and Diseases

Anthracnose (Scorch) *Colletotrichum gleosporoides*

Symptoms

Small, dark brown spots or black lesions on leaves which may be surrounded by a chlorotic halo; leaf necrosis; dieback of stem; withered leaves and scorched appearance

Cause: Fungus

Comments

Disease overwinters in plant debris; occurs worldwide

Management

The most effective method of controlling the disease is to plant yam varieties that are resistant to anthracnose such as TDA 291 or TDA 297

Mealybugs: *Rastrococcus* spp.

Symptoms

Flattened oval to round disc-like insect covered in waxy substance on tree branches; insects attract ants which may also be present; insect colony may also be associated with growth of sooty mold due to fungal colonization of sugary honeydew excreted by the insect

Cause: Insect

Comments

Insects have a wide host range; often tended by ants which farm them for their sugary honeydew secretions

Management

Prune out heavily infested branches; mealybugs can potentially be controlled by natural enemies; chemical pesticides may decrease populations of natural enemies leading to mealybug outbreaks; horticultural oils or soapy solutions can be used to treat heavy infestations.

Drumsticks

Moringa oleifera is the most widely cultivated species of the genus *Moringa*, which is the only genus in the family Moringaceae. English common names include: moringa, drumstick tree (from the appearance of the long, slender, triangular seed-pods), horseradish tree (from the taste of the roots, which resembles horseradish), ben oil tree or benzoil tree (from the oil which is derived from the seeds). It is a fast-growing, drought-resistant tree, native to the southern foothills of the Himalayas in northwestern India, and widely cultivated in tropical and sub-tropical areas where its young seed pods and leaves are used as a vegetable.

Description

Moringa oleifera *is a fast growing, evergreen, deciduous tree. It can reach a height of 10–12 m and the trunk can reach a diameter of 45 cm. The bark has a whitish-grey colour and is surrounded by thick cork. Young shoots have purplish or greenish-white hairy bark. The tree has an open crown of drooping, fragile branches and the leaves build up a feathery foliage of tripinnate leaves.*

The flowers are fragrant and bisexual, surrounded by five unequal thinly veined yellowish-white petals. The flowers are approximately 1-

1.5 cm long and 2 cm broad. They grow on slender hairy stalks in spreading or drooping later flower clusters which have a longitude of 10–25 cm.

Flowering begins within the first six months after planting. In seasonally cool regions, flowering will only occur once a year between April and June. In more constant seasonal temperature and with constant rainfall, flowering can happen twice or even all year-round. The fruit is a hanging, three-sided brown capsule of 20–45 cm size which holds dark brown, globular seeds with a diameter of approximately 1 cm. The seeds have three whitish papery wings and are dispersed by wind and water. In cultivation, it is often cut back annually to 1–2 metres and allowed to regrow so the pods and leaves remain within arm's reach.

Cultivation

The moringa tree is grown mainly in semiarid, tropical, and subtropical areas, corresponding in the United States to USDA hardiness zones 9 and 10. It grows best in dry sandy soil and tolerates poor soil, including coastal areas. As with all plants, optimum cultivation depends on producing the right environment for the plant to thrive. Moringa is a sun and heat-loving plant, and thus does not tolerate freeze or frost. Moringa is particularly suitable for dry regions, as it can be grown using rainwater without expensive irrigation techniques.

Parameter	***Requirement/Range***
Climate	Grows best in tropical or sub-tropical
Altitude/ Height	0 – 2000 metres
Rainfall	250 – 2000 mm. Irrigation needed for leaf production if rainfall < 800 mm
Soil Type	Loamy, sandy or sandy-loam
Soil pH	pH 5 - 9

Production Area

As of 2010, cultivation in Hawaii, for commercial distribution in the United States, is in its early stages. "India is the largest producer of moringa, with an annual production of 1.1 to 1.3 million tonnes of tender fruits from an area of 380 km². Among the states, Andhra Pradesh leads in both area and production (156.65 km²) followed by Karnataka (102.8 km²) and Tamil Nadu (74.08 km²). In other states, it occupies an area of 46.13 km². Tamil Nadu is the pioneering state in•so•much as it has varied genotypes from diversified geographical areas and introductions from Sri Lanka."

Moringa is grown in home gardens and as living fences in Southern India and Thailand, where it is commonly sold in local markets. In the Philippines, it is commonly grown for its leaves which are used in soup. Moringa is also actively cultivated by the World Vegetable Centre in Taiwan, a centre for vegetable research with a mission to reduce poverty and malnutrition in developing countries through improved production and consumption of vegetables. Tamil Nadu, Southern India has moringa in its folk stories and use in home gardens. In Haiti it is grown as windbreaks and to help reduce soil erosion.

Continents	*Countries Where Moringa Grows*
Africa	Benin, Burkina Faso, Cameroon, Chad, Comoros, Archipelago, Eritrea, Ethiopia, Ghana, Kenya, Madagascar, Malawi, Mali, Mauritius, Mozambique, Niger, Nigeria, Senegal, Seychelles, Sierra Leone, Somalia, Sudan, Tanzania, Togo, Zambia, Zimbabwe
Asia	Bangladesh, Burma-Myanmar, Cambodia, India, Indonesia, Laos, Malaysia, Nepal, Pakistan, Philippines, Sri Lanka, Taiwan, Timor-Leste, Thailand, Vietnam, Yemen
Latin America/Caribbean	Barbados, Belize, Brazil, Cayman Islands, Colombia, Costa Rica, Cuba, Dominican Republic, Haïti, Dutch Antilles, El Salvador, French Guiana, Guadeloupe, Guatemala, Haiti, Honduras, Mexico, Nicaragua, Panama, Panama Canal, Zone, Puerto Rico, San Salvador, Suriname, Trinidad, Venezuela, Jamaica
Oceana	Fiji, Guam, Palau

Cultivation Practice

Moringa can be grown as an annual or perennial plant. In the first year all pods are edible. Later years also bear non edible bitter pods. Therefore Moringa is often commercially cultivated annually. On less favourable locations the perennial cultivation has big advantages. Erosion is much smaller with perennial cultivation. Perennial cultivation of Moringa is also practiced in agroforestry.

Soil Preparations

In tropical cultivation sides the soil erosion is a major problem. Therefore the soil treatment has to be as low as possible. Plowing is required only for high planting densities. In low planting densities "it is better to dig pits and refill them with the soil. This ensures good root system penetration without causing too much land erosion. The pits must be 30 to 50 cm deep, and 20 to 40 cm wide."

Propagation

Moringa can be propagated from seed or cuttings. Direct seeding is possible because the germination rate of Moringa oleifera is high. After 12 days the germination rate is about 85%. Production in seedbeds or containers is very time consuming. In these technics the plants can be better protected from insects and other pests. They are also used in areas where soil erosion is a problem. Cuttings of 1 metre length and a diameter of at least 4 cm can be also used for propagation. At least one third of the cutting must be buried in the soil. In the Philippines, moringa is propagated by planting 1–2 m-long limbs cuttings, preferably from June to August. It can also be propagated by seeds, which are planted an inch below the surface and can be germinated year-round in well-draining soil.

Planting

For intensive leaf production "the spacing of plants should be 15 x 15 cm or 20 x 10 cm, with conveniently spaced alleys (for example: every 4 metres) to facilitate plantation management and harvests. Another option is to space the seeding lines 45 cm apart and to sow every 5cm on those lines. One can also space the lines only 30 cm apart and sow at a larger distance on the lines (10 to 20 cm)". Weeding and disease prevention are difficult because of the high density. In a semi-intensive production the plants are spaced 50 cm to 1 m apart. This gives good results with less maintenance. Moringa trees can also be cultivated in alleys, as natural fences and associated with other crops. The distance between moringa rows in an agroforestry cultivation are usually between 2 to 4 metres. In Haiti it is being used as fencing and windbreaks on farms.

Breeding

In India, from which Moringa most likely originates, the diversity of Moringa in cultivars in wild types is large. This gives a good basis for breeding programs. In countries where Moringa has been introduced as a cultivar, the diversity is usually much smaller among the cultivar types. Locally well adapted wild types on the other hand, can be found in most regions. Because Moringa is cultivated and used in different ways, exist different breeding aims. The breeding aims for an annual or a perennial plant are obviously different. The yield stability of fruits are an important breeding aim for the commercial cultivation in India where Moringa is cultivated annually.

On less favourable locations the perennial cultivation has big advantages. Erosion is much smaller with perennial cultivation. Perennial cultivation of Moringa is also used in agroforestry. In Pakistan varieties have been tested for their nutritional composition of the leaves on different locations. The different breeding aims result in a different selection. India selects for a higher number of pods and dwarf or semi-dwarf varieties. Breeders in Tanzania on the other hand are selecting for a higher oil content. In total, only little breeding has been achieved so far.

Yield and Harvest

Moringa oleifera can be cultivated for its leaves, pods and/or its kernels for oil extraction and water purification. The yields vary widely, depending on season, variety, fertilization, and irrigation regime. Moringa yields best under warm, dry conditions with some supplemental fertilizer and irrigation. Moringa harvest is done manually with knifes, sickles and stabs with hooks attached to it.

Fruits

When the plant is grown up from cuttings the first harvest can already take place after 6-8 month after plantation. Often, the fruits are not yielded in the first year and the yield is generally low during the first years. By year 2 it produces around 300 pods, by year 3 around 400-500. A good tree can yield 1000 or more pods. In India a hectare can produce 31 tons of pods per year. Under North Indian conditions the fruits ripen during the summer. Sometimes, particularly in South India, flowers and fruits appear twice a year and so there are 2 harvests, in July to September and March to April.

Leaves

Average yields of 6 tons/ha/year in fresh matter are can be achieved. The harvest differs strongly between the rainy and dry season with 1120 kg/ha per harvest and 690 kg/ha per harvest. The leaves and stems can be harvested from the young plants 60 days after seeding and then another 7 times in the year. At every harvest the plants are cut back to within 60 cm of the ground. In some production systems the leaves are harvested every 2 weeks. Foidl. et al. (2001) showed that the cultivation of Moringa oleifera can also be done intensively with irrigation and fertilization with suitable varieties. Trials in Nicaragua with 1 million plant/ha and 9 cuttings/year over 4 years gave an average fresh matter production of 580 metric tons per ha/year equivalent to about 174 metric tons of fresh leaves.

Oil

One estimate for yield of oil from kernels is 250 liters per hectare. The oil can be used as a food supplement, as a base for cosmetics and for hair and the skin.

Nutritional Compounds

Many parts of the moringa are edible. Regional uses of the moringa as food vary widely, and include:

- The immature seed pods, called "drumsticks", popular in Asia and Africa.
- Leaves, particularly in the Cambodia, Philippines, South India, Sri Lanka and Africa.
- Mature seeds
- Oil pressed from the mature seeds
- Roots

In some regions, the young seed pods are most commonly eaten, while in others, the leaves are the most commonly used part of the plant. The flowers are edible when cooked and are said to taste like mushrooms. The bark, sap, roots, leaves, seeds, oil, and flowers are used in traditional medicine in several countries. In Jamaica, the sap is used for a blue dye.

Leaves

The leaves are the most nutritious part of the plant, being a significant source of B vitamins, vitamin C, provitamin A as beta-carotene, vitamin K, manganese and protein, among other essential nutrients. When compared with common foods particularly high in certain nutrients per 100 g fresh weight, cooked moringa leaves are considerable sources of these same nutrients.

Nutrients	*Common Food*		*Moringa Leaves*
Vitamin A as beta-carotene	Carrot	8.3 mg	17.6 mg
Calcium	Milk	300 mg	2185 mg
Potassium	Banana	358 mg	1236 mg
Protein	Yogurt	8 g	29.4 g
Vitamin C	Orange	53 mg	52 mg

Some of the calcium in moringa leaves is bound as crystals of calcium oxalate though at levels 25-45 times less than that found spinach, which is a negligible amount.

The leaves are cooked and used like spinach. In addition to being used fresh as a substitute for spinach, its leaves are commonly dried and crushed into a powder used in soups and sauces. As with most foods, heating moringa above 140 degrees Fahrenheit destroys some of the nutritional value.

Drumsticks

The immature seed pods, called "drumsticks", are commonly consumed in South Asia. They are prepared by parboiling, and cooked in a curry until soft. The seed pods/fruits, even when cooked by boiling, remain particularly high in vitamin C (which may be degraded variably by cooking) and are also a good source of dietary fibre, potassium, magnesium and manganese.

Seeds

The seeds, sometimes removed from more mature pods and eaten like peas or roasted like nuts, contain high levels of vitamin C and moderate amounts of B vitamins and dietary minerals (right table, USDA).

Seed Oil

Mature seeds yield 38–40% edible oil called ben oil from its high concentration of behenic acid. The refined oil is clear and odorless, and resists rancidity. The seed cake remaining after oil extraction may be used as a fertilizer or as a flocculent to purify water. Moringa seed oil also has potential for use as a biofuel.

Roots

The roots are shredded and used as a condiment in the same way as horseradish; however, they contain an alkaloid, potentially having nerve-paralyzing properties.

Malnutrition Relief

Moringa trees have been used to combat malnutrition, especially among infants and nursing mothers. Five NGOs in particular — Trees for Life International, The Christian and Missionary Alliance, Church World Service, Educational Concerns for Hunger Organisation, and Volunteer Partnerships for West Africa — have advocated moringa as "natural nutrition for the tropics." One author stated that "the nutritional properties of Moringa are now so well known that there seems to be little doubt of the substantial health benefit to be realised by consumption of Moringa leaf powder in situations where starvation is imminent."

Moringa is especially promising as a food source in the tropics because the tree is in full leaf at the end of the dry season when other foods are typically scarce.

Culinary Uses

Moringa has numerous applications in cooking throughout its regional distribution. It may be preserved by canning and exported.

In Bangladesh, it is made into a variety of curry dishes by mixing with coconut, poppy seeds, and mustard or boiled until the drumsticks are semisoft and consumed directly without any extra processing or cooking. It is used in curries, sambars, kormas, and dals, although it is also used to add flavour to cutlets and other recipes. The fruit meat of drum sticks, including young seeds, is used for soup. Young leaves can either be fried with shrimp or added as a topping in fish soup.

There are several traditional Cambodian dishes using leaves (sluc) of the moringa tree known as daum m'rum, such as korko (a mixed vegetable soup). As it is a favourite vegetable, Cambodians traditionally grow moringa trees close to their residences.

In South India, Sri Lanka and Java, it is used to prepare a variety of *sambar*, is fried, or made into curry dishes by mixing with coconut, poppy seeds, and mustard or boiled until the drumsticks are semisoft and consumed directly without any extra processing or cooking. It is used in curries, sambars, kormas, and dals, although it is also used to add flavours, such as in ghee and soups. In Maharashtra, the pods are used in sweet and sour curries. In Gujarat and Rajasthan, the pods are used in to cook a spicy curry. Tender drumstick leaves, finely chopped, are used as garnish for vegetable dishes and salads. It is also used in place of or along with coriander. In some regions, the flowers are gathered and cleansed to be cooked with *besan* to make *pakoras*.

The leaves may be fried and mixed with dried-fried tuna chips (Maldive fish), onions and dried chillies. This is equivalent to a *sambal* and eaten along with rice and curry. In one area in the Maldives, a soup is made with these leaves and rice, and eaten especially for breakfast during the month of Ramazan. It is also a common ingredient in an omelet. The pods are used to cook a mild curry.

Figure: *Traditional Thai* kaeng som *with drumstick pods and fresh* pla thu

In the Punjab region of India and Pakistan, moringa called Soanjhna flowers are first separated from the stem, boiled, mashed and cooked. Curdle is an important element of its recipe to create a specific taste and favourite dish. The green pods, the leaves and the flowers are used in a variety of Thai dishes, such as curries, stir-fries, soups, omelets and salads. One of the most traditional dishes is sour Thai curry made with the drumstick pods and fish.

In the Philippines, moringa leaves, known as *kamunggay*, *malunggay* or *marungay*, are commonly added to broth as a simple soup. The leaves may also be used as a typical ingredient in *tinola*, a traditional chicken dish consisting of chicken in a broth, moringa leaves, and either green papaya or another vegetable or in the all vegetable dish known as utan. The leaves can also be processed with olive oil and salt for a pesto-like pasta sauce that has become popular on the Filipino culinary scene. Moringa juice may be mixed with lemonsito juice to make ice candies or cold drinks, possibly more palatable to those who dislike vegetables.

In 2007, Filipino Senator Loren Legarda campaigned for the popularization of moringa. She asked the government to make moringa among its priority crops for propagation, citing a Bureau of Plant Industry report about moringa's nutritional content. The leaves may also be used in making *polvoron* (a milky, powdered snack), biofuel, and ben oil.

Research

Moringa has shown a variety of potential medicinal effects in laboratory experiments, but there is no evidence supporting its use in people, except some weak evidence suggesting it might have small effect on lipid profiles. Moringa extracts are used in some primitive water filtration systems.

Other Uses

In developing countries, moringa has the potential to improve nutrition, boost food security, foster rural development, and support sustainable landcare. It may be used as forage for livestock, a micronutrient liquid, a natural anthelmintic and possible adjuvant.

Moringa has been used in folk medicine, including Siddha medicine and Ayurvedic traditional medicines and in the Philippines. In Ayurvedic traditional medicine, the leaves are believed to affect blood pressure and glucose levels. In Africa, Indonesia and Philippines, moringa leaves are given to nursing mothers in the belief that they increase lactation.